Anomaly Detection in Random Heterogeneous Media

Martin Simon

Anomaly Detection in Random Heterogeneous Media

Feynman-Kac Formulae, Stochastic Homogenization and Statistical Inversion

With a Foreword by Prof. Dr. Lassi Päivärinta

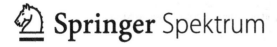 Springer Spektrum

Martin Simon
Mainz, Germany

Dissertation, Johannes Gutenberg University of Mainz, Germany, 2014

ISBN 978-3-658-10992-9   ISBN 978-3-658-10993-6 (eBook)
DOI 10.1007/978-3-658-10993-6

Library of Congress Control Number: 2015945820

Springer Spektrum
© Springer Fachmedien Wiesbaden 2015

Printed on acid-free paper

Springer Spektrum is a brand of Springer Fachmedien Wiesbaden
Springer Fachmedien Wiesbaden is part of Springer Science+Business Media
(www.springer.com)

To my parents Andrea and Alfred, my wife Lilian and my daughter Luisa.

Foreword

Inverse problems constitute an interdisciplinary field of science concentrating on the mathematical theory and practical interpretation of indirect measurements. Possible applications include medical imaging, atmospheric remote sensing, industrial process monitoring and astronomical imaging. Innovations such as computerized tomography, magnetic resonance imaging and exploration of the interior of the earth by using earthquake data are typical examples where mathematical research has played a major role. The common feature of all these problems is their extreme sensitivity to measurement noise. Dealing with this so-called *ill-posedness* both theoretically and practically often requires genuine scientific progress, say in geometry, stochastics or analysis. Through mathematical modeling it becomes in turn possible to bring these theoretical inventions to real life applications.

Martin Simon's book is a multi-field work, mainly in the fields of applied probability theory and statistical inverse problems. The work is motivated by an important inverse problem, namely electrical impedance tomography (EIT) which is also called Calderón's inverse conductivity problem. The author is interested in practical real-life applications. Therefore, he uses the so-called complete electrode model, the most accurate forward model for real-life EIT. For the same reason, he also relies on the theory of statistical inverse problems, i.e., the effort to formulate real-world inverse problems as Bayesian statistical estimation problems. That is, solutions are specified as posterior distributions and regularizing prior distributions are optimally defined as function-valued random fields which allow for fast numerical approximations, e.g., via Gaussian Markov random field approximations. Although the Bayesian framework is well-known in mathematical statistics, a rigorous application to physical problems has only been possible after solving fundamental mathematical problems regarding the existence of conditional distributions for the random variables used to describe physical fields and time-dependent variables, in other words, to solve the problem in infinite dimension. Specific to practical applications is on the other hand the need to discretize the problems in order to enable numerical computations as well as the need to deal with approximation errors related to the discretization. Moreover, independence of the latter is essential to guarantee that the calculated posterior has a consistent limit if the discretization mesh for computations is made increasingly finer and finer. While the existence of conditional distributions for the (generalized) function-valued processes was published already in 1989 [101], practical discretization-invariant priors are quite recent findings, see, e.g., Lasanen [96, 97, 98], Piiroinen [136] and Roininen et al. [142]. Also, if the direct theory itself requires, say, finite element method computations, practical ways to deal with modeling errors as well as the application of Markov Chain Monte Carlo (MCMC) methods are recent developments. Even more recent is the idea to understand deviations of the target from a deterministic model as homogenization of a stochastic random field in the study of model errors. The present monograph is a culmination in this long development, fully solving a function-valued nonlinear inverse problem with the direct problem specified by a PDE, so that both the discretization error and also small target noise are properly handled. When starting the research towards such a framework in Finland in the 1980s, this was only a distant dream, which has now come true.

In the present monograph, forward and inverse problems for EIT are approached with stochastic methods. This seems natural as the inverse problem considered is the stochastic anomaly detection problem: *Given one realization of the random current-to-voltage map corresponding to one particular realization of a random background conductivity, perturbed by a perfectly conducting anomaly occupying the region Σ, and given the distribution of the random background conductivity, locate Σ.* The author approaches the problem with stochastic homogenization theory, which he develops by utilizing a Feynman-Kac type formula for the forward problem for the conductivity equation subject to different boundary conditions. As the title indicates, the book is divided into three major parts. Chapter 2 is centered around reflected diffusion processes corresponding to the conductivity equation with merely measurable anisotropic conductivities. This chapter contains many useful new results, Theorem 2.12 and Theorem 2.15 being the most important ones. These results provide the Feynman-Kac representations (i.e. functionals of the diffusion processes) for the solutions of the continuum model and the complete electrode model, respectively. One should emphasize that these generalizations of previously known results allow extending the stochastic framework to measurable, bounded, uniformly elliptic anisotropic conductivities on domains with Lipschitz boundaries. Chapter 3 justifies the stochastic homogenization in a bounded measurement domain (namely, in a half ball with electrodes on the reflecting flat boundary and with killing on the hemisphere) in contrast to the whole space case that omits the boundary effects. Theorem 3.2 shows that for a.e. realization the potentials of the stochastic forward problem and the corresponding electrode currents converge to the potential and electrode current, respectively, of the homogenized and deterministic forward problem. The proof uses very recent invariance principles obtained by Chen, Croydon and Kumagai [33] and the Feynman-Kac representation obtained in Chapter 2. The stochastic homogenization result motivates the author to introduce a new numerical method, called continuum microscale Monte Carlo method, for estimating the effective conductivity. In Theorem 3.7 the classical argument of Kipnis and Varadhan [86] is generalized to give a quantitative result of the convergence speed for the proposed numerical method. The proof uses various techniques ranging from spectral calculus for self-adjoint operators to environment seen by the particle processes. The chapter ends with numerical computations that exploit the inherent parallelism of the proposed method. The author shows that the method is competitive and in some cases even outperforms current state-of-the-art homogenization methods. The results of Chapter 2 as well as the homogenization and convergence result from Chapter 3 follow a joint work with Piiroinen [137]. In Chapter 4, the stochastic anomaly detection problem is formulated as a statistical inverse problem and then considered numerically with Bayesian statistical methods. Here, the key aspect is the fact that the author uses the approximation error approach of Kaipio and Somersalo [78, 79] to control the random homogenization error. Finally, the numerical reconstructions for the stochastic anomaly detection problem, which are of similar quality as those for the deterministic problem, are presented.

Summarizing, it can be said that this monograph forms a piece of deep mathematical research which gives a sophisticated and elegant solution to an important and long-outstanding inverse problem.

Prof. Dr. Lassi Päivärinta

Preface

Over the last 15 years or so statistical and stochastic methods in inverse problems research have rapidly emerged, mainly as a result of the dramatic increase in computing power available to scientists and engineers. In contrast to the classical deterministic methods, the statistical approach allows to objectively determine confidence levels in the numerical reconstruction of the quantity of interest. In other words, the framework of statistical and stochastic inverse problems yields a rigorous way to characterize the impact of variability and lack-of-knowledge in the underlying mathematical models. This highly interdisciplinary field of research is at the boundary between analysis, probability theory and numerical mathematics with broad applications ranging from problems in engineering and medical or environmental imaging to problems in quantitative finance.

In this thesis, we study a stochastic extension of the classical deterministic inverse problem of EIT, also known as Calderón's problem, which is commonly used as a prototype problem. We develop both the theoretical and computational framework for the numerical solution of an inverse anomaly detection problem in heterogeneous random media. Although the details of the proposed method depend on the problem at hand, the principle idea presented in this work is applicable to a wide variety of inverse problems. The book is meant for researchers, practitioners and graduate students working in the field of statistical and stochastic inverse problems.

My research activities, which led to the writing of this monograph, have greatly benefited from the influence of numerous colleagues and friends: First and foremost, I would like to express my sincere gratitude to my "Doktorvater" Prof. Dr. Martin Hanke-Bourgeois for introducing me to the field of inverse problems and for his guidance, confidence and constant support over the last years. His dedication and attention to detail have always been a source of inspiration to me.

I am deeply grateful to Prof. Dr. Lassi Päivärinta, who, with gracious hospitality, permitted me to pursue parts of this work during two research stays at the University of Helsinki. He kindly agreed not only to act as a referee for this thesis but also to write the foreword of the monograph, which is greatly appreciated.

I wish to thank Prof. Dr. Matthias Birkner for agreeing to act as a referee for this thesis and for his detailed and insightful comments on the manuscript.

I am greatly indebted to Dr. Petteri Piiroinen for collaborating with me on the topic of this work. He kindly shared his deep mathematical understanding and I feel honored by his friendship.

I am grateful to Prof. Dr. Sylvain Maire for valuable advice, as well as numerous fruitful discussions about the topic of this work. His kind invitations to the Université de Toulon and to the INRIA Sophia Antipolis are gratefully acknowledged.

I am indebted to Prof. Dr. Nuutti Hyvönen for kindly inviting me to the Finnish Inverse Days in 2011 and for numerous thought-provoking discussions on stochastic and statistical inverse problems.

I am thankful to Prof. Dr. Nicole Marheineke for valuable advice during the first year of my doctoral studies.

I am appreciative of the fact that Prof. Dr. Elton P. Hsu took the time for both carefully reading parts of the manuscript and an inspiring discussion when he visited the Mathematics Institute at the University of Mainz in December 2013.

I am obliged to Dr. Stefanie Hollborn, Dr. Stephan Schmitz and Dr. Albrecht Seelmann for carefully reading parts of the manuscript and for valuable comments which led to improvements in the presentation of this monograph.

I wish to thank all the members of the Numerical Mathematics group at the University of Mainz, particularly our secretaries Brigitte Burkert and Jutta Gonska, for creating an excellent work environment.

I would like to thank Dr. Angelika Schulz at Springer for her friendly support and professional editorial help.

This thesis was written within the trilateral Chinese-Finnish-German research project *Inverse Problems in Electrostatics and Electrodynamics*. I gratefully acknowledge the financial support for my work from the Deutsche Forschungsgemeinschaft (DFG) and the Center for Computational Sciences Mainz (CSM).

Finally, I would like to express my heartfelt thanks to my parents, Andrea and Alfred Simon, for providing me with the kind of educational background that allowed me to pursue my goals. I dedicate this work to them as well as to my wife Lilian and my daughter Luisa, the most influential persons in my life. I am infinitely grateful for their support.

Martin Simon

Contents

Introduction

Electrical impedance tomography (EIT) aims to recover the interior electrical properties of an object from voltage and current measurements on its boundary. In practical applications, a finite number of electrodes is attached on the surface of the object and different electric current patterns are driven through them, measuring the resulting electrode voltages. Then a mathematical reconstruction algorithm is applied to the measured data to approximate the interior electrical conductivity of the object. We refer the reader to Cheney, Isaacson and Newell [36] and to Borcea [23] for review articles on the topic.

In the well-studied deterministic setting, the *forward problem* of EIT is to compute the electric potential in the object, given its conductivity and the injected boundary current. The *inverse conductivity problem*, i.e., the reconstruction of the conductivity from knowledge of the boundary potential corresponding to one or many boundary currents, is now known as *Calderón's problem* as it was posed by Alberto Calderón in his famous 1980 article [28] that laid the foundation for the mathematical study of this problem. If the full *Dirichlet-to-Neumann map*, which maps boundary potentials to the corresponding boundary currents, is given, then uniqueness of the solution to the inverse conductivity problem is known, cf., e.g., Kohn and Vogelius [91], Sylvester and Uhlmann [155], Nachman [126], Isakov [73], Astala and Päivärinta [7], Haberman and Tataru [62], Haberman [63], Caro and Rogers [30]. However, the inverse conductivity problem has also been shown to be severely ill-posed, cf. Alessandrini [3], that is, its solution is extremely sensitive to measurement and modeling errors. As a result, EIT suffers from inherent low resolution and due to this limitation, many applications focus on the detection of conductivity anomalies in a known background conductivity rather than conductivity imaging. This special case of the inverse conductivity problem is called an *inverse anomaly detection problem*.

In this thesis, we elaborate on a *stochastic* inverse anomaly detection problem involving a random heterogeneous background conductivity. In possible applications, such as geophysical subsurface imaging, the objective is to obtain information about conductivity anomalies, for instance caused by soil pollution, from electrical measurements on the surface. Indeed, it is known that the conductivity of soil contaminated by hydrocarbons is significantly increased as a result of the increment of salt content in pore water, cf. [147]. In the mathematical modeling of such problems, randomness typically reflects a lack of precise information about the meso- and microstructure of the heterogeneous ground, which may fluctuate on many scales.

As the title suggests, the work is divided into three major components. In the first part, we use the theory of symmetric Dirichlet forms to derive *Feynman-Kac formulae*, i.e., functionals of certain diffusion processes, for the solution to the EIT forward problem. More precisely, we prove these formulae for, possibly anisotropic, merely measurable conductivities and different electrode models. Subsequently, in the first chapter of the second part, we employ one of these Feynman-Kac formulae to rigorously justify *stochastic homogenization* in the case of the stochastic forward problem arising from the aforementioned anomaly detection problem. This demonstrates that on macroscopic scales, the stochastic forward problem shows an *effective* behavior. That is, the oscillations caused by the random micro-scale structure average out in an appropriate way so that the forward problem can be described by a deterministic, macroscopic forward model with a constant conductivity. Motivated by this theoretical result, we propose a new *continuum micro-scale Monte Carlo method* for the numerical approximation of the effective conductivity. The novelty of this method lies in the fact that it is based on simulation of the underlying diffusion process rather than the standard finite element discretization of the so-called *auxiliary problem*. In the second chapter of the second part, we develop a numerical method in the framework of *Bayesian inversion* for the stochastic inverse anomaly detection problem. The key ingredient of this method is an accurate approximation of the effective conductivity, obtained via the continuum micro-scale Monte Carlo method, in conjunction with the introduction of an *enhanced error model*.

In the first part of this work, we derive Feynman-Kac formulae for the solution to the deterministic forward problem. More precisely, for the solution to the *conductivity equation* for the electric potential

$$\nabla \cdot (\kappa \nabla u) = 0$$

posed on a bounded domain $D \subset \mathbb{R}^d$, $d \geq 2$, with Lipschitz boundary ∂D. κ denotes the, possibly anisotropic, uniformly bounded and uniformly elliptic conductivity and we consider different boundary conditions modeling electrode measurements. Due to these boundary conditions, the underlying diffusion processes are *reflecting*. It is well known that reflecting diffusion processes that are generated by (non-)divergence form operators with smooth coefficients are *Feller processes* which satisfy *Skorohod type* stochastic differential equations, see Lions and Sznitman [112]. However, this is not true for divergence form operators with merely measurable coefficients. Indeed, the construction of reflecting diffusion processes generated by such operators requires the theory of *symmetric Dirichlet forms* which has its origin in the *energy method* used by Dirichlet to address the boundary value problem in classical electrostatics that was subsequently named after him. Such a construction is a major challenge for an arbitrary Euclidean domain D due to the fact that the underlying Dirichlet form is not necessarily regular on $L^2(D)$, so that the reflecting diffusion process can in general only be constructed on some abstract closure of D, see Chen [32]. When D is a bounded Lipschitz domain, Bass and Hsu [15] constructed the reflecting Brownian motion living on $\overline{D}$ by showing that in this case, the so-called *Martin-Kuramochi boundary* coincides with the Euclidean boundary. A general reflecting diffusion process on a bounded Lipschitz domain, even allowing locally a finite number of Hölder cusps, was first constructed by Fukushima and Tomisaki [55]. Probabilistic approaches to both parabolic and elliptic boundary value problems for second order differential operators have been studied by many authors, starting with Feynman's Princeton thesis [48] and the article [76] by Kac. The probabilistic approach to the Dirichlet problem for a general class of second-order elliptic operators with merely measurable coefficients, even allowing singularities of a certain type, was elaborated by Chen and Zhang [34]; see also Zhang's paper [163]. However, there are only few works concerned with Feynman-Kac formulae for Neumann or Robin boundary value problems. Moreover, the approaches existing in the literature consider either the Laplacian, see, e.g., [2, 15, 27, 69], or (non-)divergence form operators with smooth coefficients, see, e.g., [51, 133, 18]. For the particular case of the conductivity equation, we generalize both the Feynman-Kac formula for the Robin boundary value problem for an isotropic $C^{2,\delta}$, $\delta > 0$, conductivity obtained by Papanicolaou [133], as well as the formula obtained by Benchérif-Madani and Pardoux [18] for the Neumann problem under similar regularity assumptions. While both of the aforementioned approaches use stochastic differential equations and Itô calculus, our approach is based on the Dirichlet form construction, following the seminal paper by Bass and Hsu [15] for the reflecting Brownian motion. Our work is related to the recent paper [35] by Chen and Zhang, where a Feynman-Kac formula for the solutions to some mixed boundary value problems with singular coefficients is derived. In contrast to our setting, however, the mixed boundary condition studiedin [35] results from a singular lower-order term of the differential operator.

The first chapter of the second part of this work is devoted to homogenization of the stochastic forward problem arising from the inverse anomaly detection problem, both theoretically as well as practically. Indeed, the homogenization theory for both periodic and random elliptic divergence form operators is well-developed, cf., e.g., [19, 131, 132, 165], but the numerical approximation of the effective conductivity in the random setting still poses major challenges. More precisely, the commonly used deterministic methods based on a discretization of the so-called *auxiliary problem* have two drawbacks: First, the auxiliary problem is formulated on the whole space $\mathbb{R}^d$ and second, it has to be solved for almost every realization of the underlying random medium. Therefore, the computational domain has to be truncated in practical computations. This raises the question about appropriate boundary conditions, cf. Bourgeat and Piatnitski [26]. Moreover, in several practically relevant cases, such as rapidly oscillating high contrast or rough random media, it is very difficult to solve the corresponding variational problems by deterministic discretization methods. More precisely, each sample produced, say, by a finite element based Monte Carlo method requires the solution of a large linear system corresponding to an extremely fine discretization of a huge computational domain in order to accurately capture the behavior of the solution. Therefore, standard numerical approximations of the effective conductivity are likely to be both prohibitively expensive and heavily biased and as a matter of fact, practitioners often choose to avoid these computations at all and rather content themselves with theoretical bounds, cf., e.g., [158]. On the other hand, it has been observed in the physics literature that the shortcomings of standard deterministic methods can be circumvented by using *continuum micro-scale Monte Carlo* simulation of certain diffusion processes evolving in the random medium instead, cf., e.g., [82, 149, 159, 84, 85]. In this work, we give a rigorous mathematical justification for homogenizing the stochastic EIT forward problem using such a method by studying the interconnection between homogenization of the stochastic boundary value problem and homogenization of the underlying diffusion process. More precisely, we employ the Feynman-Kac formula to prove a homogenization result for the corresponding stochastic boundary value problem. Homogenization of stochastic differential equations with reflection and partial differential equations with Neumann boundary conditions, respectively, have been studied for periodic coefficients in [13, 19, 156] and for random divergence form operators with smooth coefficients in [141]. Our homogenization result is valid for, possibly anisotropic, merely measurable conductivities thanks to an invariance principle for the corresponding processes which has been proved recently by Chen, Croydon and Kumagai [32]. The proof is constructive in the sense that it motivates the *numerical stochastic homogenization* based on simulation of the underlying diffusion process. Note that in contrast to solutions to stochastic differential equations with smooth coefficients, which can be simulated very efficiently, see, e.g., Kloeden and Platen [89], the diffusion processes we are interested in here evolve in discontinuous (digitized) media. In the last decade, the numerical simulation of such processes has been the subject of active research, cf., e.g., [25, 45, 46, 47, 106, 108, 110, 120, 122]. While the one-dimensional situation is now well understood, the question whether there exist exact simulation schemes for the multi-dimensional case is still unsettled. In this work, we study the special case of digitized media. The proposed simulation method uses a scheme introduced by Lejay and Maire [108] based on exact simulation of excursions of the one-dimensional *skew Brownian motion*. We provide a convergence analysis for the corresponding continuum micro-scale Monte Carlo method by generalizing a classical argument due to Kipnis and Varadhan [86]. We refer the reader to the paper [124] by Mourrat which initiated the idea of using the Kipnis and Varadhan argument in order to obtain quantitative results. Our proof is also inspired by the recent papers by Egloffe, Gloria, Mourrat and Nguyen [43] and by Gloria and Mourrat [56], which study the discrete lattice random walk in random environment. The key ingredient is a spectral estimate obtained by Gloria, Neukamm and Otto [57] in the discrete case, respectively Gloria and Otto [58] in the continuum case. The main advantage of the presented approach to numerical homogenization, beside its inherent parallelism, lies in the facts that its convergence rate is dimension-independent, its computational cost grows only linearly with the dimension and moreover, it is relatively robust with respect to the spatial regularity of the realizations of the underlying random field.

In the second chapter of the second part, we build upon the previous results in order to derive a numerical reconstruction method for the stochastic inverse anomaly detection problem. To be precise, we study the problem of detecting a parameterized anomaly in an isotropic, stationary and ergodic conductivity random field whose realizations are rapidly oscillating. On the one hand, one can of course not hope to recover all the details about the random background medium from EIT measurements because of the ill-posedness of the inverse problem. On the other hand, when it comes to locating anomalies in strongly heterogeneous media, the simplifying modeling assumption of a homogeneous background may be too restrictive due to large modeling errors. We address this ambiguity here by considering the following stochastic anomaly detection problem: *Given one realization of the random current-to-voltage map corresponding to one particular realization of a random background conductivity, perturbed by a perfectly conducting anomaly occupying the region Σ, and given the distribution of the random background conductivity, locate Σ.* For the practical solution of this problem, we propose a two-stage numerical method in the framework of Bayesian inverse problems. Numerical homogenization of the stochastic forward problem using the continuum micro-scale Monte Carlo method developed in the second part is a key ingredient of this method. The novelty of our approach lies in the introduction of an enhanced error model accounting for the approximation errors that result from reducing the full forward model to a homogenized one. That is, we aim to recover the anomaly by homogenizing the stochastic boundary value problem while still accounting for the fine-scale oscillations through the underlying error model. The method consists of a first stage *maximum a posteriori (MAP)* estimate for the reduced forward model equipped with the enhanced error model, combined with *Markov chain Monte Carlo (MCMC) bootstrap sampling* of the resulting posterior distribution in the second stage. As a result, the proposed method is not only capable of locating the anomaly, but also assesses the reliability of the estimated parameter values. In the case of a homogeneous background conductivity, such a two-stage method has been proposed by Pursiainen, cf. [139]. The basic principle of the proposed method is the *approximation error modeling* technique due to Kaipio and Somersalo, cf. [78, 79]. That is, we construct a statistical model describing both the modeling error due to homogenization as well as the approximation errors caused by truncation of the computational domain and coarse finite element discretization. While in the context of EIT and optical tomography, the approximation error approach has been successfully applied to cope with the latter, cf., e.g., [6, 100, 128, 139], the model reduction by using homogenized equations in conjunction with the statistics of the resulting *homogenization error* is a novel aspect. The Monte Carlo sampling which has to be carried out in order to precompute these statistics is, however, computationally expensive since resolving the highly heterogeneous microstructure requires an extremely fine finite element discretization. Although the details of the proposed method depend on the problem at hand, the principle idea presented in this work is applicable to a wide variety of inverse problems involving random media. The ideas presented in the third part of this work are motivated by Nolen and Papanicolaou's paper [130], where, in the case of Schrödinger operators with random potentials, closed-form expressions were employed in order to quantify both the uncertainty due to fine scale fluctuations in the coefficients of the forward problem as well as its propagation to the inverse parameter identification problem. More precisely, they modeled the effect on the solution to the forward problem using a central limit theorem from [49]. Unfortunately, analogous theoretical results for divergence form operators are only available in one spatial dimension at this point. Therefore, in contrast to their work, we have to quantify the statistics of the homogenization error numerically using Monte Carlo sampling. Our method is also somewhat related to the ideas developed in the article [11] by Banks and Criner, where homogenization theory for randomly perforated domains was used to develop a method for damage detection in porous materials. However, in contrast to our approach, the authors of [11] did not incorporate the error caused by homogenizing the forward model into the underlying error model.

Outline:

We start in Chapter 1 by introducing the basic mathematical setting of this work. Then, in Chapter 2, we describe the construction of reflecting diffusion processes via Dirichlet form theory and derive the Feynman-Kac formulae for the deterministic forward problems. Moreover, we provide Skorohod decompositions of the underlying reflecting diffusion process for two practically relevant classes of conductivities. In Chapter 3, we establish a rigorous link between homogenization of the stochastic forward problem for the complete electrode model and homogenization of the underlying diffusion process. Subsequently, we introduce a numerical method based on this theoretical result. We discuss path simulation in digitized random media, provide a convergence analysis for the proposed method and present numerical examples to support our findings. Finally, in Chapter 4 we introduce a new error model, based on homogenization of the stochastic problem, in the framework of Bayesian inverse problems. This yields a two-stage numerical reconstruction method for the stochastic anomaly detection problem. We provide numerical experiments to illustrate the feasibility of the proposed method.

In Appendix A, we recall some definitions and important results from the theory of symmetric Dirichlet forms. Appendix B is devoted to a brief description of Gaussian random field models and their simulation via the circulant embedding technique. In Appendix C, we recall the finite element discretization of the forward problem for the complete electrode model.

Part I

Probabilistic interpretation of EIT

1 Mathematical setting

In this introductory chapter, we shall briefly introduce the mathematical setting of the thesis. More precisely, we recall the well-known deterministic formulations of both the forward and the inverse problem of EIT. Afterwards, we introduce a new *stochastic* EIT anomaly detection problem which has not been studied in the literature so far.

First, we give a short presentation of our notation in the next section. Then, in Section 1.2, we recall the derivation of the deterministic EIT forward problem from the governing time-harmonic Maxwell's equations and state Calderón's inverse conductivity problem. Subsequently, we will present different electrode models from the literature with a special emphasis on the so-called *complete electrode model*. In Section 1.3, we turn to the stochastic setting. We discuss both the modeling of random heterogeneous media as well as the stochastic inverse anomaly detection problem which we are going to elaborate on in this work.

1.1 Preliminaries

Let D denote a bounded Lipschitz domain in $\mathbb{R}^d$, $d \geq 2$, with connected complement and Lipschitz parameters (r_D, c_D), i.e., for every $x \in \partial D$ we have after rotation and translation that $\partial D \cap B(x, r_D)$ is the graph of a Lipschitz function in the first $d - 1$ coordinates with Lipschitz constant no larger than c_D and $D \cap B(x, r_D)$ lies above the graph of this function. Moreover, we set $\mathbb{R}^d_- := \{x \in \mathbb{R}^d : x \cdot \nu < 0\}$, with $\nu = e_d$ the outward unit normal on $\mathbb{R}^{d-1}$, where we identify the boundary of $\mathbb{R}^d_-$ with $\mathbb{R}^{d-1}$, with straightforward abuse of notation.

For Lipschitz domains, there exists a unique outward unit normal vector ν a.e. on ∂D so that the real Lebesgue spaces $L^p(D)$ and $L^p(\partial D)$ can be defined in the standard manner with the usual L^p norms $\|\cdot\|_p$, $p = 1, 2, \infty$. The standard L^2 inner-products are denoted by $\langle \cdot, \cdot \rangle$ and $\langle \cdot, \cdot \rangle_{\partial D}$, respectively. The d-dimensional Lebesgue measure is denoted by m and the $(d - 1)$-dimensional Lebesgue surface measure is denoted by σ and $|\cdot|$ denotes the Euclidean norm on $\mathbb{R}^d$.

By $(\Gamma, \mathcal{G}, \mathcal{P})$ we always mean a complete probability space corresponding to a random medium. We use the notation ω for an arbitrary element of Γ and $\mathbb{M}$ for the expectation with respect to the probability measure $\mathcal{P}$. We use bold letters to denote functions on $(\Gamma, \mathcal{G}, \mathcal{P})$, while we use italic letters for the corresponding realizations on $\mathbb{R}^d \times \Gamma$. The canonical probability space corresponding to diffusion processes evolving in a deterministic medium starting in x is denoted $(\Omega, \mathcal{F}, \mathbb{P}_x)$ and the expectation with respect to $\mathbb{P}_x$ is denoted $\mathbb{E}_x$. If the process is evolving in a random medium, we indicate this with a superscript ω for the probability measure, i.e., the measure $\mathbb{P}_x^\omega$ corresponds to the particular realization ω of the medium. Finally, the product probability space corresponding to the *annealed measure* $\overline{\mathbb{P}} := \mathcal{P} \, \mathbb{P}_0^\omega$ on $\overline{\Omega} := \Gamma \times \Omega$, which is obtained by integrating with respect to the measure $\mathbb{P}_0^\omega$ and subsequent averaging over the realizations of the random medium, is denoted $(\overline{\Omega}, \overline{\mathcal{F}}, \overline{\mathbb{P}})$. The expectation with respect to $\overline{\mathbb{P}}$ is denoted $\overline{\mathbb{E}}$.

All functions in this work will be real-valued and derivatives are understood in distributional sense. We use a diamond subscript to denote subspaces of the standard Sobolev spaces containing functions with vanishing mean and interpret integrals over ∂D as dual evaluations with a constant function, if necessary. For example, we will frequently use the spaces

$$H_\diamond^{\pm 1/2}(\partial D) := \left\{ \phi \in H^{\pm 1/2}(\partial D) : \langle \phi, 1 \rangle_{\partial D} = 0 \right\}$$

and

$$H_\diamond^1(D) := \left\{ \phi \in H^1(D) : \langle \phi, 1 \rangle = 0 \right\}.$$

Throughout this work we mean by an *anomaly* a perturbation of the background conductivity by a perfectly conducting inclusion. We always assume that this anomaly occupies a region $\Sigma \subset D$ given by a bounded Lipschitz domain with connected complement. Moreover, we will frequently assume that ∂D is partitioned into two disjoint parts, $\partial_1 D$ and $\partial_2 D$. We denote by $H_0^1(D \backslash \Sigma \cup \partial_1 D)$ the closure of $C_c^\infty(D \backslash \Sigma \cup \partial_1 D)$, the linear subspace of $C^\infty(\overline{D} \backslash \Sigma)$ consisting of functions ϕ such that $\mathrm{supp}(\phi)$ is a compact subset of $D \backslash \Sigma \cup \partial_1 D$, in $H^1(D \backslash \Sigma)$. Furthermore, we define the Sobolev space

$$H_{0,\Sigma}^1(D \backslash \Sigma \cup \partial_1 D) := \{ \phi \in H_0^1(D \backslash \Sigma \cup \partial_1 D) : \phi = \mathrm{const} \text{ on } \partial \Sigma \}$$

and the corresponding Bochner space $L^2(\Gamma; H_{0,\Sigma}^1(D \backslash \Sigma \cup \partial_1 D))$ given by

$$\left\{ \phi : \Gamma \to H_{0,\Sigma}^1(D \backslash \Sigma \cup \partial_1 D) : \int_\Gamma \|\phi(\cdot, \omega)\|_{H_{0,\Sigma}^1(D \backslash \Sigma \cup \partial_1 D)}^2 \, \mathrm{d}\mathcal{P}(\omega) < \infty \right\},$$

see, e.g., [10] for properties of this space.

For the reason of notational compactness, we use the Iverson brackets: Let S be a mathematical statement, then

$$[S] = \begin{cases} 1, & \text{if } S \text{ is true} \\ 0, & \text{otherwise.} \end{cases}$$

We also use the Iverson brackets $[x \in B]$ to denote the indicator function of a set B, which we abbreviate by $[B]$ if there is no danger of confusion.

In what follows, all unimportant constants are denoted c, sometimes with additional subscripts, and they may vary from line to line.

1.2 Electrical impedance tomography

Consider the time-harmonic Maxwell's equations, to be precise, Faraday's law and Ampère's law

$$\text{curl} E = i\omega\mu H, \quad \text{curl} H = -(i\omega\varepsilon - \kappa)E, \tag{1.1}$$

where ω is the frequency, μ the *magnetic permeability*, ε the *electrical permittivity* and κ the *electrical conductivity*. The physically relevant electric and magnetic field are given by the real parts $\text{Re}(E(x)e^{i\omega t})$ and $\text{Re}(H(x)e^{i\omega t})$, respectively. Let us consider the *quasi-static* case, i.e., the case of low frequencies ω, then the imaginary part of the *electrical admittivity* $i\omega\varepsilon - \kappa$ becomes negligible as well as the term $i\omega\mu H$. In fact, it can be shown that the Maxwell system is approximated by

$$\text{curl} E = 0, \quad \text{curl} H = \kappa E, \tag{1.2}$$

see, e.g., [36]. In this case, the electric field must be a gradient field $E = -\nabla u$ for the scalar *electric potential* u. Substitution of this expression into the second equation in (1.2) and taking the divergence finally yields the elliptic *conductivity equation*

$$\nabla \cdot (\kappa \nabla u) = 0 \quad \text{in } D. \tag{1.3}$$

This approximation is highly accurate for low-frequency time-harmonic (sinusoidal alternating input currents) or static (direct input currents) potentials. In this work, we restrict ourselves to the latter case in which the electrical conductivity κ is real-valued. More precisely, we assume that the, possibly anisotropic, conductivity $\kappa \in L^{\infty}(D; \mathbb{R}^{d \times d})$ can be represented by a symmetric matrix-valued function $\kappa : D \to \mathbb{R}^{d \times d}$, $x \mapsto (\kappa_{ij}(x))_{i,j=1}^{d}$ with components in $L^{\infty}(D)$. Moreover, we always assume that κ is uniformly bounded and uniformly elliptic, i.e., there exists some constant $c > 0$ such that

$$c^{-1}|\xi|^2 \leq \xi \cdot \kappa(x)\xi \leq c|\xi|^2, \quad \text{for every } \xi \in \mathbb{R}^d \text{ and a.e. } x \in D. \qquad (1.4)$$

1.2.1 Calderón's problem

In the so-called *continuum model*, the conductivity equation (1.3) is equipped with a co-normal boundary condition

$$\partial_{\kappa\nu}u := \kappa\nu \cdot \nabla u|_{\partial D} = f \quad \text{on } \partial D, \qquad (1.5)$$

where f is a measurable function modeling the signed density of the outgoing current. The boundary value problem (1.3), (1.5) has a solution if and only if

$$\langle f, 1 \rangle_{\partial D} = 0. \qquad (1.6)$$

Physically speaking, this means that the current must be conserved. Given an appropriate function f, the solution to (1.3), (1.5) is unique up to an additive constant, which physically corresponds to the choice of the ground level of the potential. If $f \in H_\diamond^{-1/2}(\partial D)$, then there exists a unique equivalence class of functions $u \in H^1(D)/\mathbb{R}$ that satisfies the weak formulation of the boundary value problem

$$\int_D \kappa \nabla u \cdot \nabla v \, dx = \langle f, v|_{\partial D} \rangle_{\partial D} \quad \text{for all } v \in H^1(D)/\mathbb{R},$$

where $v|_{\partial D} := \gamma v$ and $\gamma : H^1(D)/\mathbb{R} \to H^{1/2}(\partial D)/\mathbb{R} = (H_\diamond^{-1/2}(\partial D))'$ is the standard trace operator. One can thus define the *Dirichlet-to-Neumann map*

$$\Lambda_\kappa : H_\diamond^{1/2}(\partial D) \to H_\diamond^{-1/2}(\partial D), \quad \phi \mapsto \partial_{\kappa\nu}u. \qquad (1.7)$$

Note that we occasionally write v instead of $v|_{\partial D}$ for the sake of readability.

The inverse conductivity problem for the continuum model, commonly known as *Calderón's problem*, reads as follows: *Given the Dirichlet-to-Neumann map Λ_κ (or its inverse), is it possible to determine the conductivity κ uniquely? If yes, how can one reconstruct (features of) κ from a possibly noisy, incomplete version of Λ_κ?*

In two dimensions and for isotropic $\kappa \in L_+^\infty(D)$, the question of uniqueness was answered affirmatively in the celebrated paper [7] by Astala and Päivärinta. The recent result by Caro and Rogers [30] yields the same assertion for $d \geq 3$, provided that κ is a Lipschitz function. When the conductivity is not assumed to be isotropic, there is always an obstruction to uniqueness, namely we have $\Lambda_{\kappa_1} = \Lambda_{\kappa_2}$, whenever $\kappa_2 = F_* \kappa_1$ is the push-forward conductivity by a diffeomorphism F on D that leaves the boundary ∂D invariant. In the plane, this is known to be the only obstruction by the result of Astala, Päivärinta and Lassas [8] which holds without additional regularity assumptions on the conductivity. In higher dimensions, it is an open question whether such a result holds true, see the paper by Dos Santos Ferreira, Kenig, Salo and Uhlmann, [42] for further discussion.

Even when the full Dirichlet-to-Neumann map is given, the inverse conductivity problem is unstable. To be precise, there exist stability estimates, cf. Alessandrini [3] and Barceló, Barceló and Ruiz [12], but it is known that they can not be better than logarithmic, cf. Mandache [118]. That is, given two isotropic conductivities κ_1 and κ_2, an estimate of the form

$$||\kappa_1 - \kappa_2||_\infty \le c_1 \left| \log||\Lambda_{\kappa_1} - \Lambda_{\kappa_2}||_{H_\diamond^{1/2}(\partial D) \to H_\diamond^{-1/2}(\partial D)} \right|^{-c_2} \tag{1.8}$$

holds with positive constants c_1 and c_2; see also the stability estimates obtained by Knudsen, Lassas, Mueller and Siltanen [90] for noisy versions of the Dirichlet-to-Neumann map, which are of logarithmic type as well. Therefore, the inverse conductivity problem is in the class of *severely* or *exponentially ill-posed* inverse problems, cf. Engl, Hanke and Neubauer [44].

1.2.2 Electrode models

In practical EIT measurements, a number of electrodes, denoted $E_1, ..., E_N \subset \partial D$, are attached on the boundary of the object D, see Figure 1.1. The electrodes are modeled by disjoint surface patches given by simply connected subsets of ∂D, each having a Lipschitz boundary.

We call the vector $J^{(k)} = (J_1^{(k)}, ..., J_N^{(k)})^T \in \mathbb{R}^N$ a *current pattern*, if it satisfies the conservation of charges condition

$$\sum_{l=1}^{N} J_l^{(k)} = 0 \tag{1.9}$$

and analogously the vector $U^{(k)} = (U_1^{(k)}, ..., U_N^{(k)})^T \in \mathbb{R}^N$ is called a *voltage pattern*, if it satisfies the grounding condition

$$\sum_{l=1}^{N} U_l^{(k)} = 0. \tag{1.10}$$

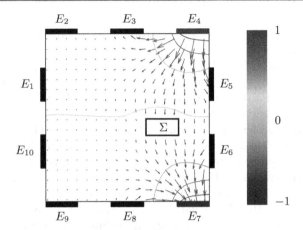

Figure 1.1: Current density $\kappa\nabla u$ (arrows) and equipotential lines; imposed voltage pattern: $U = (0,0,0,1,0,0,-1,0,0,0)^T$. The subdomain Σ models a perfectly conducting anomaly in unit background conductivity.

Commonly, $K \leq N - 1$ linearly independent current patterns $J^{(k)}$, $k = 1, ..., K$, are injected through the electrodes and the resulting voltage patterns $U^{(k)}$, $k = 1, ..., K$, are measured. Note that we will suppress the superscript of the current and voltage patterns whenever this causes no confusion.

A simple but rather crude electrode model is the so-called *gap model*, where it is assumed that the electrodes have positive surface measure and that they are perfect conductors. By the latter assumption, the current density must be constant on each electrode. Moreover, it is assumed that the boundary ∂D is insulating apart from the electrodes, i.e., the current density through $\partial D \backslash \{E_1, ..., E_N\}$ vanishes. The *point electrode model* is a related model, where the electrodes are modeled as point current sources such that the boundary condition (1.5) holds in the sense of distributions. The relation between the point electrode model and the gap model is studied in the appendix of the paper [66] by Hanke, Hyvönen and Reusswig. Note moreover, that the gap model could be viewed as a discretization of the Neumann-to-Dirichlet map for a certain basis of current and voltage patterns.

One can observe from physical experiments that the current density through an electrode is not constant in practice. This shunting effect of the highly conductive electrodes is accounted for by the *shunt model*, where the potential is assumed to be constant on each electrode.

The most accurate forward model for real-life EIT is the so-called *complete electrode model* (CEM) which takes into account all of the aforementioned physical effects plus the fact that during electrode measurements there is a *contact impedance* caused by a thin, highly resistive layer at the electrode object interface. It has been demonstrated experimentally that the complete electrode model can correctly predict EIT measurements up to instrument precision, cf. [150]. For a given current pattern $J \in \mathbb{R}^N$ the boundary conditions for the electric potential in the complete electrode model are given by

$$\int_{E_l} \partial_{\kappa\nu} u \, d\sigma(x) = J_l, \quad 1 \le l \le N, \tag{1.11}$$

together with the Robin boundary condition

$$\kappa\nu \cdot \nabla u|_{\partial D} + gu|_{\partial D} = f \quad \text{on } \partial D, \tag{1.12}$$

where the functions $f, g : \partial D \to \mathbb{R}$ are defined by

$$f(x) := \frac{1}{z(x)} \sum_{l=1}^{N} U_l[E_l], \quad g(x) := \frac{1}{z(x)} \sum_{l=1}^{N} [E_l] \tag{1.13}$$

and the contact impedance $z : \partial D \to \mathbb{R}$ is assumed to be an integrable function satisfying

$$0 < c_0 \le z \le c_1 \quad \text{a.e. on } \partial D.$$

Note that the electrode voltage vector U in the definition of f is a priori unknown and therefore part of the solution of the forward problem. By the Lax-Milgram theorem, the condition (1.9) ensures existence and uniqueness of the forward solution in the quotient space $(H^1(D) \oplus \mathbb{R}^N)/\mathbb{R}$, cf. [150]. That is, the elements (u, U) and (v, V) are equivalent if

$$u - v = U_1 - V_1 = \ldots = U_N - V_N = \text{const}$$

and the grounding condition (1.10) picks out a unique solution.

Although it is common in practical EIT measurements to apply currents and measure the corresponding electrode voltages, note that from a theoretical point of view, we may equivalently assume that a voltage pattern U is imposed and the corresponding current pattern J is measured. Then the variational form of the boundary value problem (1.3), (1.12) reads as follows: Given $U \in \mathbb{R}^N$ satisfying (1.10), find $u \in H^1(D)$ such that

$$\int_D \kappa\nabla u\nabla v \, dx + \langle gu|_{\partial D}, v|_{\partial D}\rangle_{\partial D} = \langle f, v|_{\partial D}\rangle_{\partial D} \quad \text{for all } v \in H^1(D). \tag{1.14}$$

Knowledge of u yields the corresponding electrode currents via (1.11). Figure 1.1, which is based on numerical data simulated using a finite element method, illustrates the EIT forward problem modeled by the complete electrode model. For a thorough study of the relation between discrete electrode measurements based on the complete electrode model and idealized measurements based on the continuum model, we refer the reader to Hyvönen [71].

Assuming that it is possible to measure on the whole boundary, the electric potential u modeled by the complete electrode model satisfies the Robin boundary condition

$$z\kappa\nu \cdot \nabla u|_{\partial D} + u|_{\partial D} = h \quad \text{on } \partial D$$

so that the inverse conductivity problem is then to reconstruct κ from the knowledge of the *Robin-to-Neumann map*

$$\mathcal{R}_{z,\kappa} : h \mapsto \kappa\nu \cdot \nabla u|_{\partial D}$$

for all boundary currents $h \in H^{-1/2}(\partial D)$. It has been shown by Kolehmainen, Lassas and Ola [93], that $\mathcal{R}_{z,\kappa}$ uniquely determines the map z and hence the knowledge of $\mathcal{R}_{z,\kappa}$ is equivalent to the knowledge of the Dirichlet-to-Neumann map Λ_κ. In particular, all of the aforementioned uniqueness results for Calderón's problem apply.

1.3 A stochastic anomaly detection problem

The basic geometric setting of the stochastic problem we are interested in is as follows: Assume that the model domain is given by the lower hemisphere

$$D := B(0,R) \cap \mathbb{R}^d_-, \quad R > 0,$$

and that ∂D is partitioned into two disjoint parts, namely the *accessible boundary* $\partial_1 D := \partial D \cap \mathbb{R}^{d-1}$ and the *inaccessible boundary* $\partial_2 D := \partial D \backslash \partial_1 D$, respectively. We usually choose R large such that the corresponding measurements resemble data obtained from half-space measurements.

Such a setting is found for instance in geophysical applications, where measurements can only be taken on the surface and the perfectly conducting anomaly occupying the region Σ models, e.g., polluted ground, see Figure 1.2 for an illustration. In the mathematical modeling of such problems, randomness of the background medium typically reflects a lack of precise information about the meso- and microstructure of the heterogeneous ground, which may fluctuate on many scales.

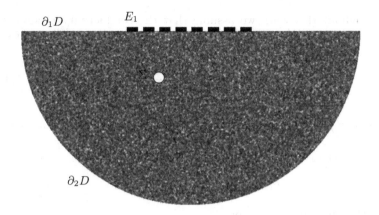

Figure 1.2: Basic geometric setting and measurement configuration with 8
electrodes; a realization of a random background conductivity is
shown and the anomaly is depicted in white.

1.3.1 Random heterogeneous media

Let $(\Gamma, \mathcal{G}, \mathcal{P})$ be a probability space and let $\Theta : \Gamma \to \Gamma$ denote an ergodic d-dimensional *dynamical system*, i.e., a family of automorphisms $\{\Theta_x, x \in \mathbb{R}^d\}$ which satisfies the following conditions:

(i) The family $\{\Theta_x, x \in \mathbb{R}^d\}$ is a group, i.e., $\Theta_0 = \mathrm{id}$ and

$$\Theta_{x+y} = \Theta_x \Theta_y \quad \text{for all } x, y \in \mathbb{R}^d;$$

(ii) the mappings $\Theta_x : \Gamma \to \Gamma$, $x \in \mathbb{R}^d$, preserve the measure $\mathcal{P}$ on Γ, i.e., for every $B \in \mathcal{G}$, $\Theta_x B$ is $\mathcal{P}$-measurable and

$$\mathcal{P}(\Theta_x B) = \mathcal{P}(B);$$

(iii) for every measurable function ϕ on $(\Gamma, \mathcal{G}, \mathcal{P})$, the function $(x, \omega) \mapsto \phi(\Theta_x \omega)$ is a measurable function on $(\mathbb{R}^d \times \Gamma, \mathcal{B}(\mathbb{R}^d) \otimes \mathcal{G}, m \times \mathcal{P})$, where $\mathcal{B}(\mathbb{R}^d) \otimes \mathcal{G}$ denotes the sigma-algebra generated by the measurable rectangles;

(iv) the family $\{\Theta_x, x \in \mathbb{R}^d\}$ is ergodic, i.e., $\phi(\Theta_x \omega) = \phi(\omega)$ for all $x \in \mathbb{R}^d$ and $\mathcal{P}$-a.e. $\omega \in \Gamma$ implies $\phi = \mathrm{const}$ $\mathcal{P}$-a.e.

Throughout this work we assume that the conductivity random field $\{\kappa(x,\omega), (x,\omega) \in \mathbb{R}^d \times \Gamma\}$ is the *stationary extension* on $\mathbb{R}^d \times \Gamma$ of some function $\kappa \in L^2(\Gamma; \mathbb{R}^{d \times d})$, that is,

$$(x,\omega) \mapsto \kappa(x,\omega) = \kappa(\Theta_x \omega). \tag{1.15}$$

Note that if κ can be written in the form (1.15) with a dynamical system $\{\Theta_x, x \in \mathbb{R}^d\}$ which satisfies (i)-(iii), then it is automatically *stationary* with respect to $\mathcal{P}$, i.e., for every finite collection of points $x^{(i)}$, $i = 1, ..., k$, and any $h \in \mathbb{R}^d$ the joint distribution of

$$\kappa(x^{(1)} + h, \omega), ..., \kappa(x^{(k)} + h, \omega)$$

under $\mathcal{P}$ is the same as that of

$$\kappa(x^{(1)}, \omega), ..., \kappa(x^{(k)}, \omega).$$

Even if it is not explicitly stated, we always assume that the conductivity random field may be written in the form (1.15), where the underlying dynamical system $\{\Theta_x, x \in \mathbb{R}^d\}$ satisfies conditions (i)-(iv). Moreover, we will explicitly state if we assume that the conductivity random field $\{\kappa(x,\omega), (x,\omega) \in \mathbb{R}^d \times \Gamma\}$ satisfies one of the following assumptions:

(A1) $\kappa \in L^2(\Gamma; L^\infty(\mathbb{R}^d; \mathbb{R}^{d \times d}))$ and the random field is strictly positive and uniformly bounded, that is, there exists a constant $c > 0$ such that for every $\xi \in \mathbb{R}^d$ and a.e. $x \in \mathbb{R}^d$

$$\mathcal{P}\left\{\omega \in \Gamma : c^{-1}|\xi|^2 \leq \xi \cdot \kappa(x,\omega)\xi \leq c|\xi|^2\right\} = 1.$$

(A2) $\{\kappa(x,\omega), (x,\omega) \in \mathbb{R}^d \times \Gamma\}$ satisfies the *spectral gap property*, cf. [58]: There exist constants $\rho, r > 0$ such that for all measurable functions on $\left\{\kappa : \mathbb{R}^d \to \{\kappa_0 \in \mathbb{R}^{d \times d} : |\kappa_0 \xi| \leq |\xi|, c|\xi|^2 \leq \xi \cdot \kappa_0 \xi \text{ for all } \xi \in \mathbb{R}^d\}\right\}$

$$\mathbb{V}\phi \leq \frac{1}{\rho}\mathbb{M}\int_{\mathbb{R}^d} \left(\operatorname{osc}_{\kappa|_{B(x,r)}} \phi\right)^2 dx,$$

where we have set

$$\left(\operatorname{osc}_{\kappa|_{B(x,r)}} \phi\right) := \sup\left\{\phi(\tilde\kappa) : \tilde\kappa \in \Omega, \tilde\kappa|_{\mathbb{R}^d \setminus B(x,r)} = \kappa|_{\mathbb{R}^d \setminus B(x,r)}\right\}$$

$$- \inf\left\{\phi(\tilde\kappa) : \tilde\kappa \in \Omega, \tilde\kappa|_{\mathbb{R}^d \setminus B(x,r)} = \kappa|_{\mathbb{R}^d \setminus B(x,r)}\right\}.$$

To account for the highly heterogeneous properties of the background medium, the latter is modeled using the conductivity random field with appropriate scaling by a small parameter $\varepsilon > 0$, i.e.,

$$\kappa_\varepsilon(\cdot, \cdot) : \mathbb{R}^d \times \Gamma \to \mathbb{R}^{d \times d}, \quad \kappa_\varepsilon(x, \omega) := \kappa(x/\varepsilon, \omega).$$

If the correlation length of the conductivity random field κ is, say 1, then the correlation length of the scaled version κ_ε is of order ε and for $\varepsilon \ll 1$ we obtain thus a rapidly oscillating random field.

1.3.2 Stochastic forward and inverse problem

Before we can formulate the stochastic inverse anomaly detection problem which we elaborate on in this work, let us first introduce a suitable stochastic forward problem based on the complete electrode model: We search for a random field $\{u_\varepsilon(x, \omega), (x, \omega) \in \overline{D} \times \Gamma\}$ with $\boldsymbol{u}_\varepsilon \in L^2(\Gamma; H^1_{0, \Sigma}(D \backslash \Sigma \cup \partial_1 D))$ such that the stochastic conductivity equation

$$\nabla \cdot (\kappa_\varepsilon \nabla u_\varepsilon) = 0 \quad \text{in } D \backslash \Sigma \times \Gamma \tag{1.16}$$

subject to the boundary conditions

$$\begin{aligned}
\kappa_\varepsilon \nu \cdot \nabla u_\varepsilon|_{\partial_1 D} + g u_\varepsilon|_{\partial_1 D} &= f && \text{on } \partial_1 D \times \Gamma \\
u_\varepsilon|_{\partial_2 D} &= 0 && \text{on } \partial_2 D \times \Gamma \\
u_\varepsilon|_{\partial \Sigma} &= \text{const} && \text{on } \partial \Sigma \times \Gamma
\end{aligned} \tag{1.17}$$

is satisfied for $\mathcal{P}$-a.e. $\omega \in \Gamma$. The variational formulation of the forward problem is to find $\boldsymbol{u}_\varepsilon \in L^2(\Gamma; H^1_{0, \Sigma}(D \backslash \Sigma \cup \partial_1 D))$ such that

$$\mathbb{M}\left\{ \int_{D \backslash \Sigma} \kappa_\varepsilon \nabla u_\varepsilon \cdot \nabla v \, dx + \langle g u_\varepsilon, v \rangle_{\partial_1 D} \right\} = \mathbb{M} \langle f, v \rangle_{\partial_1 D} \tag{1.18}$$

for all $v \in L^2(\Gamma; H^1_{0, \Sigma}(D \backslash \Sigma \cup \partial_1 D))$. For a given voltage pattern $U \in \mathbb{R}^N$, the corresponding measurement data is given by the random current pattern $J(\varepsilon, \omega) = (J_1(\varepsilon, \omega), ..., J_N(\varepsilon, \omega))^T$, defined for $\mathcal{P}$-a.e. $\omega \in \Gamma$ by

$$J_l(\varepsilon, \omega) = \frac{1}{|E_l|} \int_{E_l} \kappa_\varepsilon(x, \omega) \nu \cdot \nabla u_\varepsilon(x, \omega)|_{\partial_1 D} \, d\sigma(x), \quad l = 1, ..., N. \tag{1.19}$$

Note that the constant potential prescribed on $\partial \Sigma$ in (1.17) is implicitly defined by the imposed voltage pattern, the particular realization of the random background conductivity and the conservation of charges condition

$$\sum_{l=1}^N J_l(\varepsilon, \omega) = 0.$$

Both the well-posedness of the variational formulation (1.18) and the well-posedness of the stochastic forward problem (1.16), (1.17) follow from a straightforward application of the Lax-Milgram theorem. Moreover, standard arguments from measure theory show that the solution to the stochastic forward problem (1.16), (1.17) also solves the variational problem (1.18) and vice versa, cf. [10]. We refer the reader to the recent publications [64, 102] by Hakula, Hyvönen and Leinonen, where a related stochastic forward problem for the complete electrode model is studied from the point of view of stochastic Galerkin methods.

Let us denote the random finite-dimensional linear current-to-voltage map by $R_{z,\kappa_\varepsilon|_{D\setminus\Sigma}}$, that is,

$$R_{z,\kappa_\varepsilon|_{D\setminus\Sigma}}U^{(k)} = J^{(k)}(\varepsilon,\omega), \quad k = 1,...,N-1.$$

As in the deterministic case, we may equivalently apply deterministic electrode currents and consider the corresponding random finite-dimensional voltage-to-current map

$$R^{-1}_{z,\kappa_\varepsilon|_{D\setminus\Sigma}}J^{(k)} = U^{(k)}(\varepsilon,\omega), \quad k = 1,...,N-1.$$

The stochastic forward problem (1.16), (1.17) yields *partial measurements* confined to the accessible boundary $\partial_1 D$, approximating electrode measurements in the half-space. A deterministic EIT anomaly detection problem based on the continuum model in the half-space was studied in [68] by Hanke and Schappel. In particular, [68, Theorem 5.1] yields an identifiability result given measurements supported on some bounded subset of $\mathbb{R}^{d-1}$ which motivates us to formulate the following *stochastic inverse anomaly detection problem*: *Given one particular realization of the voltage-to-current map $R_{z,\kappa_\varepsilon(\cdot,\omega_0)|_{D\setminus\Sigma}}$ (or its inverse) corresponding to one particular realization $\kappa_\varepsilon(\cdot,\omega_0)$ of the background conductivity, perturbed by an anomaly occupying the region $\Sigma \subset D$, and given the distribution of the random background conductivity $\{\kappa_\varepsilon(x,\omega), (x,\omega) \in \mathbb{R}^d \times \Gamma\}$, find the region Σ.*

2 Feynman-Kac formulae

In this chapter, we establish the connection between the *deterministic* EIT forward problem and the class of *reflecting diffusion processes*. We proceed along the lines of the recent paper [137] by Piiroinen and the author: We derive *Feynman-Kac formulae* in terms of these processes for the solutions to the forward problems corresponding to the continuum model and the complete electrode model, respectively. These results extend the classical Feynman-Kac formulae for elliptic boundary value problems in smooth domains and with smooth coefficients which were obtained in the 1980s and 1990s using the Feller semigroup approach and Itô stochastic calculus. In contrast to this well-studied situation, the underlying reflecting diffusion processes in this work are constructed via Dirichlet form theory, which has emerged as a powerful tool when it comes to studying boundary value problems with non-smooth coefficients, boundaries and data.

However, in general such a construction is not very convenient with regard to practical issues such as numerical simulation. Unlike the Feller semigroup approach, which uses a pointwise analysis, the Dirichlet form approach is based on *quasi-sure* analysis, implying that we are permitted to ignore certain *exceptional sets* which are not visited by the process. Therefore, processes constructed via Dirichlet form theory are in general only defined for *quasi-every* starting point in the state space, rather than for every starting point. Moreover, processes generated by divergence form operators with merely measurable coefficients do in general not belong to the class of solutions to stochastic differential equations. Rather, they can be decomposed into a local martingale and an abstract additive functional with finite quadratic variation but possibly infinite variation; or equivalently, they admit a decomposition into two processes which are semimartingales with respect to two different filtrations. Both decompositions involve processes which are only implicitly defined and are therefore not suited for numerical simulation. It will turn out, that these issues can be resolved at least in practically relevant special cases of the EIT forward problem.

We start in Section 2.1 by constructing the underlying reflecting diffusion processes via Dirichlet form theory. We show that the transition kernel densities of these processes are Hölder continuous up to the boundary so that we may *refine* the processes to start from every point in the state space. In Section 2.2, we provide so-called *Skorohod decompositions*, i.e., semimartingale decompositions of the reflecting diffusion processes for two practically relevant classes of conductivities, thus enabling efficient numerical simulation. Section 2.3, the derivation of the Feynman-Kac formulae, is the central part of this chapter.

2.1 Reflecting diffusion processes

In his seminal paper [52], Fukushima established a one-to-one correspondence between regular symmetric Dirichlet forms and symmetric Hunt processes which is the foundation for the construction of stochastic processes via Dirichlet form techniques. Therefore, we assume that the reader is familiar with the theory of symmetric Dirichlet forms, as elaborated for instance in the monographs [54, 115]. A concise collection of both terminology as well as fundamental results is provided in Appendix A.

Let us consider the measure space $(\overline{D}, \mathcal{B}(\overline{D}), [D] \cdot m)$ as well as the following symmetric bilinear form on $L^2(D)$:

$$\mathcal{E}(v, w) := \int_D \kappa \nabla v(x) \cdot \nabla w(x) \, dx, \quad v, w \in \mathcal{D}(\mathcal{E}) := H^1(D). \tag{2.1}$$

For the particular case $\kappa \equiv 1/2$, which is of special importance, we set

$$\mathcal{E}^{\mathrm{BM}}(v, w) := \frac{1}{2} \int_D \nabla v(x) \cdot \nabla w(x) \, dx, \quad v, w \in \mathcal{D}(\mathcal{E}^{\mathrm{BM}}) := H^1(D). \tag{2.2}$$

Proposition 2.1. *The pair $(\mathcal{E}, \mathcal{D}(\mathcal{E}))$ defined by (2.1) is a strongly local regular symmetric Dirichlet form on $L^2(D)$.*

Proof. First, we verify that $(\mathcal{E}, \mathcal{D}(\mathcal{E}))$ is a symmetric Dirichlet form on $L^2(D)$. Closedness is obvious since for all $v \in H^1(D)$ we can find positive constants c_1, c_2 such that

$$c_1 \|v\|_2^2 \leq \mathcal{E}_1(v, v) \leq c_2 \|v\|_2^2.$$

This follows from (1.4). To show that the unit contraction operates on $(\mathcal{E}, \mathcal{D}(\mathcal{E}))$, we follow [54, Example 1.2.1, Example 1.2.3], i.e., we construct for each $\varepsilon > 0$ a differentiable function $\phi_\varepsilon : \mathbb{R} \to \mathbb{R}$ such that $\phi_\varepsilon(t) = t$ for all $t \in [0, 1]$, $-\varepsilon \leq \phi_\varepsilon(t) \leq 1 + \varepsilon$ for all $t \in \mathbb{R}$ and $0 \leq \phi_\varepsilon(s) - \phi_\varepsilon(t) \leq s - t$, whenever $t < s$. Given such a function we have for every $v \in H^1(D)$ that $\phi_\varepsilon(v) \in H^1(D)$ and

$$\mathcal{E}(\phi_\varepsilon(v), \phi_\varepsilon(v)) = \int_D |\phi_\varepsilon'(v(x))|^2 \kappa \nabla v \cdot \nabla v \, dx \leq \mathcal{E}(v, v),$$

where the last inequality is a consequence of the property $0 \leq \phi'_\varepsilon(t) \leq 1$. As $\mathcal{E}$ is closed, this is equivalent to the fact that the unit contraction operates on $(\mathcal{E}, \mathcal{D}(\mathcal{E}))$, cf. [54]. The function ϕ_ε can be constructed by the following standard technique: We consider the mollifier

$$
\rho(x) = \begin{cases} c \cdot \exp(-(1 - |x|^2)^{-1}), & |x| < 1 \\ 0, & |x| \geq 1, \end{cases}
$$

where the constant c is such that $\int_{|x|<1} \rho(x)\,\mathrm{d}x = 1$. Moreover, we set $\rho_\delta(x) := \delta^{-1}\rho(\delta^{-1}x)$, $\delta > 0$, $\psi_\varepsilon(t) := ((-\varepsilon) \vee t) \wedge (1 + \varepsilon)$ and define for $0 < \delta < \varepsilon$ the function

$$
\phi_\varepsilon(t) := \rho_\delta * \psi_\varepsilon(t) = \int_\mathbb{R} \rho_\delta(t - s)\psi_\varepsilon(s)\,\mathrm{d}s.
$$

The strong local property is obvious and $(\mathcal{E}, \mathcal{D}(\mathcal{E}))$ is regular by the fact that $H^1(D) \cap C(\overline{D})$ is dense in both $C(\overline{D})$ equipped with the uniform norm, as well as $H^1(D)$ equipped with the standard Sobolev norm. $\square$

Due to Proposition 2.1 and Theorem A.24, there exist an $\mathcal{E}$-exceptional set $\mathcal{N} \subset \overline{D}$ and a conservative diffusion process $X = (\Omega, \mathcal{F}, \{X_t, t \geq 0\}, \mathbb{P}_x)$, starting from $x \in \overline{D}\backslash\mathcal{N}$. X is associated with $(\mathcal{E}, \mathcal{D}(\mathcal{E}))$ in the sense of Theorem A.24. That is, for every non-negative Borel function ϕ, the transition semigroup of X defined by

$$
P_t\phi(x) := \mathbb{E}_x\phi(X_t), \quad x \in \overline{D}\backslash\mathcal{N}, \tag{2.3}
$$

is a version of the strongly continuous sub-Markovian contraction semigroup $T_t\phi$ on $L^2(D)$ associated with $(\mathcal{E}, \mathcal{D}(\mathcal{E}))$. Without loss of generality let us assume that X is defined on the *canonical sample space* $\Omega = C([0, \infty); \overline{D})$. It is well known that the symmetric Hunt process associated with (2.2) is the *reflecting Brownian motion*. In analogy to this terminology, we call the symmetric Hunt process associated with (2.1) a *reflecting diffusion process*.

Let us briefly recall the concept of the *boundary local time* of reflecting diffusion processes, see, e.g., [69, 133, 18], which will be crucial for the subsequent derivation of the Feynman-Kac formulae. If the diffusion process is the solution to a stochastic differential equation, say the reflecting Brownian motion, then the boundary local time is given by the one-dimensional process L in the Skorohod decomposition, which prevents the sample paths from leaving $\overline{D}$, i.e.,

$$
X_t = x + W_t - \frac{1}{2}\int_0^t \nu(X_s)\,\mathrm{d}L_s, \tag{2.4}
$$

$\mathbb{P}_x$-a.s. for q.e. $x \in \overline{D}$, where W is a standard d-dimensional Brownian motion. This boundary local time is a continuous non-decreasing process which increases only when $X_t \in \partial D$, namely for all $t \geq 0$ and q.e. $x \in \overline{D}$

$$L_t = \int_0^t [\partial D](X_s) \, dL_s,$$

$\mathbb{P}_x$-a.s. and

$$\mathbb{E}_x \int_0^t [\partial D](X_s) \, ds = 0.$$

Although the reflecting diffusion process associated with (2.1) does in general not admit a Skorohod decomposition of the form (2.4), we may still define a continuous one-dimensional process with these properties. More precisely, by the Lipschitz property of ∂D, we have that $D \cap B(x, r_D) = \{(\tilde{x}, x_d) : x_d > \gamma(\tilde{x})\} \cap B(x, r_D)$ and the Lipschitz function γ is differentiable a.e. with a bounded gradient. In particular, we have for every Borel set $B \subset \partial D \cap B(x, r_D)$ that

$$\sigma(B) = \int_{\{\tilde{x}:(\tilde{x},\gamma(\tilde{x}))\in B\}} \left(1 + |\nabla\gamma(\tilde{x})|^2\right)^{1/2} d\tilde{x}$$

and a straightforward computation yields that the Lebesgue surface measure σ is a smooth measure with respect to $(\mathcal{E}, \mathcal{D}(\mathcal{E}))$ having finite energy, i.e.,

$$\int_{\partial D} |v| \, d\sigma(x) \leq c||v||_{\mathcal{E}_1} \quad \text{for all } v \in \mathcal{D}(\mathcal{E}) \cap C(\overline{D}).$$

Definition 2.2. The positive continuous additive functional of X whose Revuz measure is given by the Lebesgue surface measure σ on ∂D, i.e., the unique $L \in \mathcal{A}_c^+$ such that

$$\lim_{t\to 0+} \frac{1}{t} \int_D \mathbb{E}_x \left\{ \int_0^t \phi(X_s) \, dL_s \right\} \psi(x) \, dx = \int_{\partial D} \phi(x)\psi(x) \, d\sigma(x) \qquad (2.5)$$

for all non-negative Borel functions ϕ and all α-excessive functions ψ, is called the *boundary local time* of the reflecting diffusion process X.

Remark 2.3. An equivalent construction of the boundary local time, which is, however, less convenient for our purpose, goes as follows: Set

$$L_t^\varepsilon := \varepsilon^{-1} \int_0^t [D_\varepsilon](X_s) \, ds, \quad D_\varepsilon := \{x \in \overline{D} : d(x, \partial D) \leq \varepsilon\},$$

then one can show that

$$\mathbb{E}_x |L_t^\varepsilon - L_t|^2 \to 0 \quad \text{as } \varepsilon \to 0$$

uniformly in $x \in \overline{D}$. Moreover, there exists a monotonically decreasing null sequence $(\varepsilon_k, k \in \mathbb{N})$ such that

$$\lim_{k \to \infty} L_t^{\varepsilon_k} = L_t \quad \mathbb{P}_x\text{-a.s.}$$

for q.e. $x \in \overline{D}$, uniformly in t on any compact time interval. This is the analogue of the definition of the local time for one-dimensional diffusion processes by Itô and McKean, cf. [74].

The rest of this section is devoted to showing that the $\mathcal{E}$-exceptional set $\mathcal{N}$ is actually empty. Therefore, we consider the non-positive definite self-adjoint operator $(\mathcal{L}, \mathcal{D}(\mathcal{L}))$ associated with the Dirichlet form $(\mathcal{E}, \mathcal{D}(\mathcal{E}))$. That is, for $v \in \mathcal{D}(\mathcal{L})$ we have

$$\langle -\mathcal{L}v, w \rangle = \mathcal{E}(v, w) \quad \text{for all } w \in \mathcal{D}(\mathcal{E}) \tag{2.6}$$

and the domain of $\mathcal{L}$ is given by

$$\mathcal{D}(\mathcal{L}) = \left\{ v \in \mathcal{D}(\mathcal{E}) : \exists \phi \in L^2(D) \text{ s.t. } \mathcal{E}(v, w) = \int_D \phi w \, dx \; \forall w \in \mathcal{D}(\mathcal{E}) \right\}.$$

In order to *refine* the reflecting diffusion process X to start from every $x \in \overline{D}$, we exploit the connection between the strongly continuous sub-Markovian contraction semigroup $\{T_t, t \geq 0\}$ on $L^2(D)$ and the evolution system corresponding to $(\mathcal{L}, \mathcal{D}(\mathcal{L}))$, see, e.g., the monograph [134]. Namely, for every $v_0 \in L^2(D)$, the trajectory $v : (0, T) \to H^1(D)$, $v(t) = T_t v_0$ belongs to the function space

$$\{\phi \in L^2((0, T); H^1(D)) : \dot{\phi} \in L^2((0, T); H^{-1}(D))\}$$

and is the unique mild solution to the parabolic *abstract Cauchy problem*

$$\dot{v} + \mathcal{L}v = 0 \quad \text{in } (0, T)$$
$$v(0) = v_0. \tag{2.7}$$

This is equivalent to the variational formulation

$$-\int_0^T \langle v(t), w \rangle \, \dot{\varphi}(t) \, dt + \int_0^T \langle \mathcal{L}v(t), w \rangle \, \varphi(t) \, dt - \langle v_0, w \rangle \, \varphi(0) = 0 \tag{2.8}$$

for all $w \in H^1(D)$ and all $\varphi \in C_c^\infty([0, T))$. Moreover, T_t is known to be a bounded operator from $L^1(D)$ to $L^\infty(D)$ for every $t > 0$. Therefore, by the Dunford-Pettis theorem, it can be represented as an integral operator for every $t > 0$,

$$T_t\phi(x) = \int_D p(t, x, y)\phi(y)\,dy \quad \text{for every } \phi \in L^1(D), \qquad (2.9)$$

where for all $t > 0$ we have $p(t, \cdot, \cdot) \in L^\infty(D \times D)$ and $p(t, \cdot, \cdot) \geq 0$ a.e. We call the function p the *transition kernel density* of X.

The following proposition adapts a well-known result for diffusion processes on $\mathbb{R}^d$, cf. [151], which follows from the famous De Giorgi-Nash-Moser theorem, to the case of reflecting diffusion processes on $\overline{D}$. The key idea of the proof is the following *extension by reflection* technique from [160, Section 2.4.3]: We extend the solution to a parabolic problem by reflection at the boundary. Then we show that this extension again solves a parabolic problem so that we can apply the interior regularity result due to De Giorgi, Nash and Moser. See also the article [129] by Nittka, where such a technique is applied to elliptic boundary value problems.

Proposition 2.4. $p \in C^{0,\delta}((0, T] \times \overline{D} \times \overline{D})$ *for some* $\delta \in (0, 1)$, *i.e., for each fixed* $0 < t_0 \leq T$, *there exists a positive constant* c *such that*

$$|p(t_2, x_2, y_2) - p(t_1, x_1, y_1)| \leq c(\sqrt{t_2 - t_1} + |x_2 - x_1| + |y_2 - y_1|)^\delta \qquad (2.10)$$

for all $t_0 \leq t_1 \leq t_2 \leq T$ *and all* $(x_1, y_1), (x_2, y_2) \in \overline{D} \times \overline{D}$. *Moreover, the mapping* $t \mapsto p(t, \cdot, \cdot)$ *is analytic from* $(0, \infty)$ *to* $C^{0,\delta}(\overline{D} \times \overline{D})$.

Proof. First note that Nash's inequality holds for the underlying Dirichlet form $(\mathcal{E}, H^1(D))$, i.e., there exists a constant $c_1 > 0$ such that

$$||v||_2^{2+4/d} \leq c_1(\mathcal{E}(v, v) + ||v||_2^2)||v||_1^{4/d} \quad \text{for all } v \in H^1(D).$$

This is a direct consequence of the uniform ellipticity (1.4) and [15, Corollary 2.2], where Nash's inequality is shown to hold for the Dirichlet form $(\mathcal{E}^{\text{BM}}, H^1(D))$ for a bounded Lipschitz domain D. Analogously to the proof of [15, Theorem 3.1], it follows thus from [29, Theorem 3.25] that the transition kernel density satisfies an Aronson type Gaussian upper bound

$$p(t, x, y) \leq c_1 t^{-d/2} \exp\left(-\frac{|x - y|^2}{c_2 t}\right) \qquad (2.11)$$

for all $t \leq 1$ and all $(x, y) \in \overline{D} \times \overline{D}$. In particular, $\sup_{t_0 < t \leq 1} ||p(t, \cdot, \cdot)||_\infty$ is finite and hence by the interior Hölder continuity obtained from the De

Giorgi-Nash-Moser theorem, cf. [127, 151], the estimate (2.10) is true for all (x_1, y_1), (x_2, y_2) satisfying $d(x_i, \partial D), d(y_i, \partial D) > c_3$, $i = 1, 2$, for some constant $c_3 > 0$ and all $t_0 \leq t_1 \leq t_2 \leq 1$. Note that by the semigroup property the Chapman-Kolmogorov equation holds, i.e.,

$$p(t_1 + t_2, x, y) = \int_D p(t_1, x, z)\, p(t_2, z, y)\, dz \qquad (2.12)$$

for every pair $t_1, t_2 \geq 0$ and a.e. $x, y \in \overline{D}$. In particular, for fixed $y \in \overline{D}$ the function $v := p(\cdot, \cdot, y)$ is the unique solution to (2.7) with initial value $v_0 := p(0, \cdot, y) \in L^2(D)$. Now let $z \in \partial D$ so that by the Lipschitz property of ∂D we have after translation and rotation $B(z, r_D) \cap \overline{D} = \{(\tilde{x}, x_d) \in B(z, r_D) : x_d \geq \gamma(\tilde{x})\}$ and $B(z, r_D) \cap \partial D = \{\tilde{x} \in B(z, r_D) : x_d = \gamma(\tilde{x})\}$, where we have introduced the notation $\tilde{x} = (x_1, ..., x_{d-1})^T$. Let us furthermore introduce the one-to-one transformation $\Psi(x) := (\tilde{x}, x_d - \gamma(\tilde{x}))$ which straightens the boundary $B(z, r_D) \cap \partial D$. Ψ is a bi-Lipschitz transformation and the Jacobians of both Ψ and Ψ^{-1} are bounded with bounds that depend only on the Lipschitz constant c_D. Since v is the solution to (2.7) with appropriate initial condition, the function $\hat{v} := v(\cdot, \Psi^{-1}(\cdot))$ must satisfy the following variational formulation of the a parabolic problem in $\hat{D}(z, r_D) := \Psi(B(z, r_D) \cap \overline{D})$, namely

$$\int_0^T \dot{\varphi}(t) \int_{\hat{D}(z,r_D)} \hat{v}(t) w \, dx \, dt = -\sum_{i,j=1}^d \int_0^T \varphi(t) \int_{\hat{D}(z,r_D)} \hat{\kappa}_{ij} \partial_i \hat{v}(t) \partial_j w \, dx \, dt$$

$$-\varphi(0) \int_{\hat{D}(z,r_D)} \hat{v}_0 w \, dx$$

for all $w \in C_c^\infty(\hat{D}(z, r_D))$ and all $\varphi \in C_c^\infty([0, T))$. The coefficient $\hat{\kappa}$ is obtained via change of variables and it is bounded and uniformly elliptic by the boundedness of the Jacobians of Ψ and Ψ^{-1}, respectively. Now we use reflection at the hyperplane $\{(\tilde{y}, 0)\}$ via the mapping $\rho(x) := (\tilde{x}, -x_d)$ which yields that the function $\hat{v}(\cdot, \rho(\cdot))$ satisfies the variational formulation of a parabolic problem on $\rho(\hat{D}(z, r_D))$. Summing up both variational formulations on $\hat{D}(z, r_D)$ and on $\rho(\hat{D}(z, r_D))$, respectively, we obtain that the function

$$\check{v}(t, x) := \begin{cases} \hat{v}(t, x), & x \in \hat{D}(z, r_D) \\ \hat{v}(t, \rho(x)), & x \in \rho(\hat{D}(z, r_D)) \end{cases}$$

satisfies the variational formulation of a parabolic problem in $\hat{D}(z, r_D) \cup \rho(\hat{D}(z, r_D))$. By the interior Hölder estimate for $\breve{v}$, together with the fact that we may choose $c_3 = r_D/4c_D$, we obtain thus

$$|p(t_2, x_2, \Psi^{-1}(y_2)) - p(t_1, x_1, \Psi^{-1}(y_1))| \leq c_1(\sqrt{t_2 - t_1} + |y_2 - y_1|)^{c_2}$$

for all $t_0 \leq t_1 \leq t_2 \leq 1$ and $y_1, y_2 \in \{(\tilde{x}, x_d) : |\tilde{x}| < c_3, \ x_d \in (0, r_D/4)\}$. As Ψ is bi-Lipschitz, for fixed x, the mapping $(t, y) \mapsto p(t, x, y)$ is Hölder continuous in $(t_0, 1] \times (B(z, c_3) \cap \overline{D})$ and by symmetry of the transition kernel density the same holds true for the mapping $(t, x) \mapsto p(t, x, y)$ for fixed y. Finally, the first assertion on $(t_0, 1] \times \overline{D} \times \overline{D}$ follows due to compactness of ∂D and its generalization to arbitrary $T > 0$ is obtained after repeatedly applying the Chapman-Kolmogorov equation.

The second assertion follows by the fact that the semigroup $\{T_t, t \geq 0\}$ extrapolates to a holomorphic semigroup on $L^2(D)$. More precisely, the semigroup possesses a holomorphic extension to the sector $\Sigma_\theta := \{re^{i\alpha} : r > 0, |\alpha| < \theta\}$ for some $\theta \in (0, \frac{\pi}{2}]$, cf., e.g., [134]. Let $0 < t_0 \leq T$ and set

$$\Sigma_\theta(t_0, T) := \{z \in \mathbb{C} : z - t_0 \in \Sigma_\theta, |z| < T\}.$$

By the Hölder continuity of p, the set $\{p(z, \cdot, \cdot) : z \in \Sigma_\theta(t_0, T)\}$ is a bounded subset of $C^{0,\delta}(\overline{D} \times \overline{D})$. Moreover, the family of functionals obtained from integration against the functions $[B_1](x)[B_2](y)$ for measurable $B_1, B_2 \subset \overline{D}$ form a separating subspace of $(C^{0,\delta}(\overline{D}, \overline{D}))'$, i.e., for $k \in C^{0,\delta}(\overline{D} \times \overline{D})$

$$\int_{D \times D} k(x, y)[B_1](x)[B_2](y) \, dx \, dy = 0 \quad \text{for all measurable } B_1, B_2 \subset \overline{D}$$

implies that $k \equiv 0$. As the mapping

$$z \mapsto \langle T_z[B_1], [B_2] \rangle = \int_{D \times D} p(z, x, y)[B_1](y)[B_2](x) \, dx \, dy$$

is holomorphic for all $z \in \Sigma_\theta$, the mapping $z \mapsto p(z, \cdot, \cdot)$ is holomorphic from $\Sigma_\theta(t_0, T)$ to $C^{0,\delta}(\overline{D} \times \overline{D})$ by [5, Theorem 3.1]. Since t_0 and T were arbitrary, the assertion is proved. $\qquad\square$

By [53, Theorem 2], the existence of a Hölder continuous transition kernel density ensures that we may refine the process X to start from every $x \in \overline{D}$ by identifying the strongly continuous semigroup $\{T_t, t \geq 0\}$ with the transition semigroup $\{P_t, t \geq 0\}$. In particular, if v is continuous and locally in $H^1(D)$, the Fukushima decomposition holds for every $x \in \overline{D}$, i.e.,

$$v(X_t) = v(X_0) + M_t^v + N_t^v, \quad \text{for all } t > 0, \tag{2.13}$$

$\mathbb{P}_x$-a.s., where M^v is a martingale additive functional of X having finite energy and N^v is a continuous additive functional of X having zero energy. Moreover, both M^v and N^v can be taken to be additive functionals of X in the strict sense, cf. [54, Theorem 5.2.5].

Finally, note that the 1-potential of the Lebesgue surface measure σ of ∂D is the solution to an elliptic boundary value problem on a Lipschitz domain with bounded data. By elliptic regularity theory, cf., e.g., [60], this solution is continuous, implying that the boundary local time L exists as a positive continuous additive functional in the strict sense, cf. [54, Theorem 5.1.6].

2.2 Skorohod decompositions

In this section, we derive Skorohod decompositions of the reflecting diffusion process X for two practically relevant special cases, namely local Lipschitz conductivities and isotropic piecewise constant conductivities.

The assertion of the following proposition is already covered by [55, Theorem 2.3]; we include a proof for the sake of self-containedness.

Proposition 2.5. *Let $\kappa \in C^{0,1}_{loc}(\overline{D}; \mathbb{R}^{d\times d})$ be a symmetric, uniformly bounded and uniformly elliptic conductivity. Then the reflecting diffusion process X admits the following Skorohod decomposition*

$$X_t = x + \int_0^t B(X_s)\,\mathrm{d}W_s + \int_0^t \nabla\kappa(X_s)\,\mathrm{d}s - \int_0^t \kappa(X_s)\nu(X_s)\,\mathrm{d}L_s, \quad (2.14)$$

$\mathbb{P}_x$-a.s., where $B : \overline{D} \to \mathbb{R}^{d\times d}$ denotes the positive definite diffusion matrix satisfying $B^2 = 2\kappa$, W is a standard d-dimensional Brownian motion and L is the boundary local time of X.

Proof. We have shown in Section 2.1, that the Fukushima decomposition holds with a unique martingale additive functional M^v in the strict sense and a unique continuous additive functional N^v in the strict sense. Let us first compute the energy measure of M^v. For $v, w \in \mathcal{D}(\mathcal{E})$ we obtain using Lemma A.9

$$\int_D w(x)\,\mathrm{d}\mu_{\langle M^v\rangle}(x) = \lim_{t\to 0+} \frac{1}{t}\int_D \mathbb{E}_x\left\{(v(X_t) - v(x_0))^2\right\} w(x)\,\mathrm{d}x$$

$$= \lim_{t\to 0+} \frac{1}{t}\int_D (T_t v^2(x) - 2v(x)T_t v(x) + v^2(x))w(x)\,\mathrm{d}x$$

$$
\begin{aligned}
&= \lim_{t \to 0+} \frac{2}{t} \int_D v(x)w(x)(v(x) - T_t v(x))\, \mathrm{d}x \\
&\quad - \lim_{t \to 0+} \frac{1}{t} \int_D v^2(x)(w(x) - T_t w(x))\, \mathrm{d}x \\
&= 2\mathcal{E}(vw, v) - \mathcal{E}(v^2, w) \\
&= 2 \int_D \kappa \nabla v(x) \cdot \nabla v(x)\, w(x)\, \mathrm{d}x,
\end{aligned}
$$

which yields the energy measure

$$
\mathrm{d}\mu_{\langle M^v \rangle}(x) = 2 \sum_{i,j=1}^{d} \kappa_{ij}(x)\partial_i v(x)\partial_j v(x)\, \mathrm{d}x
$$

so that the predictable quadratic variation of M^v is given by

$$
\langle M^v \rangle_t = 2 \int_0^t \sum_{i,j=1}^{d} \kappa_{ij}(X_s)\partial_i v(X_s)\partial_j v(X_s)\, \mathrm{d}s. \tag{2.15}
$$

Using the coordinate mappings $\phi_i(x) := x_i$, $i = 1, ..., d$, on $\overline{D}$ yields that M^ϕ is a continuous martingale additive functional in the strict sense with covariation

$$
\langle M^{\phi_i}, M^{\phi_j} \rangle_t = 2 \int_0^t \kappa_{ij}(X_s)\, \mathrm{d}s,
$$

$\mathbb{P}_x$-a.s. A standard characterization of continuous martingales, cf., e.g., [72], yields that

$$
M_t^v = \int_0^t (B(X_s)\nabla v(X_s)) \cdot \mathrm{d}W_s, \tag{2.16}
$$

$\mathbb{P}_x$-a.s., where $B : \overline{D} \to \mathbb{R}^{d \times d}$ denotes the positive definite diffusion matrix satisfying $B^2 = 2\kappa$ and W is a standard d-dimensional Brownian motion.

Now let us consider the continuous additive functional N^v. Again using the coordinate mappings on $\overline{D}$, we obtain from Green's formula that

$$
\begin{aligned}
\mathcal{E}(\phi_i, w) &= \sum_{j=1}^{d} \int_D \kappa_{ij}(x)\partial_j w(x)\, \mathrm{d}x \\
&= -\sum_{j=1}^{d} \int_D \partial_j \kappa_{ij}(x)w(x)\, \mathrm{d}x + \sum_{j=1}^{d} \int_{\partial D} \kappa_{ij}(x)\nu_j(x)w(x)\, \mathrm{d}\sigma(x)
\end{aligned}
$$

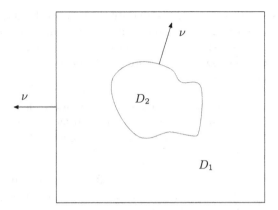

Figure 2.1: A simplistic two-phase medium.

for all $v \in H^1(D) \cap C(\overline{D})$. That is, by Lemma A.29, N^{ϕ_i} is associated with the signed Radon measure

$$-\sum_{j=1}^{d} \partial_j \kappa_{ij} \, \mathrm{d}x + \sum_{j=1}^{d} \kappa_{ij}(x)\nu_j(x) \, \mathrm{d}\sigma(x)$$

and by the fact that the unique positive continuous additive functionals in the strict sense having as Revuz measure the Lebesgue measure and the Lebesgue surface measure, respectively, are given by the constant additive functional t and the boundary local time L_t, we have shown that for every $x \in \overline{D}$

$$N_t^{\phi_i} = \sum_{j=1}^{d} \int_0^t \partial_j \kappa_{ij}(X_s) \, \mathrm{d}s - \sum_{j=1}^{d} \int_0^t \kappa_{ij}(X_s)\nu_j(X_s) \, \mathrm{d}L_s, \qquad (2.17)$$

$\mathbb{P}_x$-a.s. Substitution of (2.16) and (2.17) in the Fukushima decomposition for the coordinate mappings finally yields the Skorohod decomposition (2.14). $\qquad \square$

Now let us turn to the case of isotropic piecewise constant conductivities and for simplicity of the presentation let us consider a simplistic two-phase medium, where

$$\kappa(x) = \begin{cases} \kappa_1, & x \in D_1 \\ \kappa_2, & x \in D_2, \end{cases} \qquad (2.18)$$

with constants $\kappa_1, \kappa_2 > 0$ and D is a simply connected bounded Lipschitz domain which consists of two disjoint subdomains such that $D_1 = D \backslash \overline{D}_2$. We assume that D_2 is a simply connected Lipschitz domain. ν is the outer unit normal vector on ∂D and the outer unit normal vector on ∂D_2 with respect to D_2, see Figure 2.1.

Definition 2.6. The positive continuous additive functional L^0 of X whose Revuz measure is given by the scaled Lebesgue surface measure $(\kappa_1 + \kappa_2)\sigma$ on ∂D_2 is called the *symmetric local time* of the reflecting diffusion process X at ∂D_2.

Remark 2.7. The term "symmetric" comes from the fact that in the one-dimensional case L^0 is the local time defined by the Tanaka formula with the convention $\text{sign}(0) = 0$, which is called the symmetric local time, see [140]. In this case we have

$$L_t^0 = \lim_{\varepsilon \to 0} \frac{1}{2\varepsilon} \int_0^t [[-\varepsilon, \varepsilon]](X_s)\, \mathrm{d}s.$$

Proposition 2.8. *Let κ be given by (2.18). Then the reflecting diffusion process X admits the following Skorohod decomposition*

$$X_t = x + \int_0^t \sqrt{2\kappa(X_s)}\, \mathrm{d}W_s + \frac{\kappa_1 - \kappa_2}{\kappa_1 + \kappa_2} \int_0^t \nu(X_s)\, \mathrm{d}L_s^0 - \kappa_1 \int_0^t \nu(X_s)\, \mathrm{d}L_s,$$

$\mathbb{P}_x$-*a.s., where W is a standard d-dimensional Brownian motion, L^0 is the symmetric local time of X at ∂D_2 and L is the boundary local time.*

Proof. Repeating the computations from the proof of Proposition 2.5 yields for the martingale additive functional the predictable quadratic variation

$$\langle M^v \rangle_t = 2 \int_0^t \kappa(X_s)|\nabla v(X_s)|^2\, \mathrm{d}s, \quad \mathbb{P}_x\text{-a.s. for every } x \in \overline{D},$$

implying that

$$M_t^v = \int_0^t \sqrt{2\kappa(X_s)}\nabla v(X_s) \cdot \mathrm{d}W_s, \quad \mathbb{P}_x\text{-a.s. for every } x \in \overline{D},$$

where W is a standard d-dimensional Brownian motion.

By Green's formula we have for all $w \in \mathcal{D}(\mathcal{E}) \cap C(\overline{D})$

$$
\begin{aligned}
\mathcal{E}(v, w) &= \kappa_1 \int_{D_1} \nabla v \cdot \nabla w \, dx + \kappa_2 \int_{D_2} \nabla v \cdot \nabla w \, dx \\
&= -\int_D w \kappa \Delta v \, dx - (\kappa_1 - \kappa_2) \int_{\partial D_2} \partial_\nu v w \, d\sigma(x) \\
&\quad + \kappa_1 \int_{\partial D} \partial_\nu v w \, d\sigma(x).
\end{aligned}
$$

Using the coordinate mappings on $\overline{D}$ we obtain by Lemma A.29 that N^{ϕ_i} is associated with the signed Radon measure

$$
-(\kappa_1 - \kappa_2)\nu(x) \, d\sigma|_{\partial D_2}(x) + \kappa_1 \nu(x) \, d\sigma|_{\partial D}(x).
$$

The assertion follows by the fact that the unique positive continuous additive functionals in the strict sense having as Revuz measure the (scaled) Lebesgue surface measure on ∂D_2 and ∂D, respectively, are given by the symmetric local time L^0 at ∂D_2 and the boundary local time L_t. □

2.3 Derivation of the Feynman-Kac formulae

In this section, we derive the Feynman-Kac formulae for both the continuum model and the complete electrode model. Afterwards, we will obtain, as a corollary, a Feynman-Kac formula for the mixed boundary value problem corresponding to the stochastic anomaly detection problem introduced in the first chapter. Compared to the earlier works [69, 133, 18] on Feynman-Kac formulae, the main difficulty in deriving these formulae in our particular setting comes from the lack of Itô's formula for general reflecting diffusion processes.

The rest of this subsection is devoted to providing some auxiliary lemmata.

Lemma 2.9. *The transition kernel density p approaches the stationary distribution uniformly and exponentially fast, that is, there exist positive constants c_1 and c_2 such that for all $(x, y) \in \overline{D} \times \overline{D}$ and every $t \geq 0$,*

$$
|p(t, x, y) - |D|^{-1}| \leq c_1 \exp(-c_2 t). \tag{2.19}
$$

Proof. Fix $x \in \overline{D}$. Then by the Chapman-Kolmogorov equation

$$
\begin{aligned}
p(t,x,x) - |D|^{-1} &= \int_D p(t/2,x,y)p(t/2,y,x)\,\mathrm{d}y - |D|^{-1} \\
&= \int_D (p(t/2,x,y))^2\,\mathrm{d}y - |D|^{-1} \\
&= \int_D (p(t/2,x,y) - |D|^{-1})^2\,\mathrm{d}y \geq 0,
\end{aligned}
$$

where we have used that $\int_D p(t,x,y)\,\mathrm{d}y = 1$ for all $t \geq 0$ and $x \in \overline{D}$. Moreover, we have by the analyticity of the mapping $t \mapsto p(t,x,x)$, cf. Proposition 2.4, and the fact that p solves a parabolic boundary value problem with homogeneous Neumann boundary condition

$$
\begin{aligned}
\frac{\mathrm{d}}{\mathrm{d}t}(p(t,x,x) - |D|^{-1}) &= \frac{\mathrm{d}}{\mathrm{d}t}\int_D p(t/2,x,y)p(t/2,y,x)\,\mathrm{d}y \\
&= \frac{\mathrm{d}}{\mathrm{d}t}\int_D (p(t/2,x,y))^2\,\mathrm{d}y \\
&= -\int_D \kappa(y)\nabla_y p(t/2,x,y) \cdot \nabla_y p(t/2,x,y)\,\mathrm{d}y \\
&\leq -c^{-1}\int_D \nabla_y p(t/2,x,y) \cdot \nabla_y p(t/2,x,y)\,\mathrm{d}y \\
&\leq -c^{-1}c_D^{-1}\left(\int_D (p(t/2,x,y))^2\,\mathrm{d}y - |D|^{-1}\right) \\
&= -c^{-1}c_D^{-1}\left(p(t,x,x) - |D|^{-1}\right),
\end{aligned}
$$

where we have used (1.4) and the Poincaré inequality

$$
\left\|\phi - |D|^{-1}\int_D \phi(x)\,\mathrm{d}x\right\|_2 \leq c_D\|\nabla\phi\|_2 \quad \text{for all } \phi \in H^1(D).
$$

Integration of the inequality from above yields a diagonal estimate, i.e., there exist positive constants c_1 and c_2 such that

$$
0 \leq p(t,x,x) - |D|^{-1} \leq c_1\exp(-c_2 t) \quad \text{for every } t \geq 0.
$$

Now, the assertion follows from the Cauchy-Schwarz inequality and the fact that, by the computations from above, we may write the expression $|p(t,x,y) - |D|^{-1}|$ in the form

$$\left| \int_D (p(t/2,x,z) - |D|^{-1})(p(t/2,z,y) - |D|^{-1}) \, dz \right|$$

$$\leq \left(\int_D (p(t/2,x,z) - |D|^{-1})^2 \, dz \right)^{1/2} \left(\int_D (p(t/2,y,z) - |D|^{-1})^2 \, dz \right)^{1/2}$$

$$= \left(p(t/2,x,x) - |D|^{-1} \right)^{1/2} \left(p(t/2,y,y) - |D|^{-1} \right)^{1/2}$$

$\square$

Lemma 2.10. *Let* $\kappa \in C^\infty(\overline{D}; \mathbb{R}^{d \times d})$. *Then the set*

$$V_\kappa(D) := \{\phi : \phi \in C^2(D), \partial_{\kappa\nu}\phi = 0 \text{ a.e. on } \partial D\} \cap H^1(D) \qquad (2.20)$$

is dense in $H^1(D)$.

Proof. Diagonalizing the operator $(\mathcal{L}, \mathcal{D}(\mathcal{L}))$ corresponding to the conductivity κ, we obtain an orthonormal basis $\{\phi_k, k \in \mathbb{N}\}$ of $L^2(D)$ and an increasing sequence $(\lambda_k)_{k \in \mathbb{N}}$ of real positive numbers which tend to infinity such that for every $k \in \mathbb{N}$, $\phi_k \in H^1(D)$ is a weak solution to the corresponding eigenvalue problem with homogeneous Neumann boundary condition. Note that the inner product $\mathcal{E}_1(\cdot, \cdot)$ is equivalent to the standard inner product on $H^1(D)$. Moreover, $V_\kappa(D)$ contains the linear span of $\{\phi_k, k \in \mathbb{N}\}$ by elliptic regularity theory so it is enough to show that the linear span of eigenfunctions is dense. Therefore, let $\psi \in H^1(D)$ such that $\mathcal{E}_1(\phi_k, \psi) = 0$ for every $k \in \mathbb{N}$, then

$$0 = \int_D \kappa \nabla \phi_k \cdot \nabla \psi \, dx + \langle \phi_k, \psi \rangle = (\lambda_k + 1) \langle \phi_k, \psi \rangle.$$

Hence it follows $\langle \phi_k, \psi \rangle = 0$ for every $k \in \mathbb{N}$ and the fact that $\{\phi_k, k \in \mathbb{N}\}$ is an orthonormal basis of $L^2(D)$ implies $\psi \equiv 0$ which proves the assertion. $\square$

Lemma 2.11. *For every* $x \in \overline{D}$ *and every bounded Borel function* ϕ *on* ∂D *we have*

$$\mathbb{E}_x \int_0^t \phi(X_s) \, dL_s = \int_0^t \int_{\partial D} p(s,x,y)\phi(y) \, d\sigma(y) \, ds \quad \text{for all } t \geq 0. \quad (2.21)$$

Proof. First, note that the expression (2.21) is well-defined since the boundary local time of X exists as a positive continuous additive functional in the strict sense. Without loss of generality we may assume that ϕ is non-negative. It follows from [54, Theorem 5.1.3] that the Revuz correspondence (2.5) is equivalent to

$$\int_D \psi(x)\, \mathbb{E}_x \int_0^t \phi(X_s)\, \mathrm{d}L_s\, \mathrm{d}x \;=\; \int_0^t \int_{\partial D} \phi(y) T_s \psi(y)\, \mathrm{d}\sigma(y)\, \mathrm{d}s$$

$$=\; \int_D \psi(x) \int_0^t \int_{\partial D} p(s,y,x)\phi(y)\, \mathrm{d}\sigma(y)\, \mathrm{d}s\, \mathrm{d}x$$

for every $t > 0$ and all non-negative Borel functions ψ and ϕ. In particular, as this holds for every non-negative Borel function ψ, we deduce

$$\mathbb{E}_x \int_0^t \phi(X_s)\, \mathrm{d}L_s = \int_0^t \int_{\partial D} p(s,x,y)\phi(y)\, \mathrm{d}\sigma(y)\, \mathrm{d}s \quad \text{a.e. in } \overline{D}.$$

To obtain the assertion everywhere in $\overline{D}$, fix an arbitrary $x_0 \in \overline{D}$ and consider for $t_0 > 0$ the integral

$$\mathbb{E}_{x_0} \int_{t_0}^t \phi(X_s)\, \mathrm{d}L_s \;=\; \int_D p(t_0, x_0, x)\, \mathbb{E}_x \int_0^{t-t_0} \phi(X_s)\, \mathrm{d}L_s\, \mathrm{d}x$$

$$=\; \int_D p(t_0, x_0, x) \left(\int_0^{t-t_0} \int_{\partial D} p(s,x,y)\phi(y)\, \mathrm{d}\sigma(y)\, \mathrm{d}s \right) \mathrm{d}x$$

$$=\; \int_{t_0}^t \int_{\partial D} p(s, x_0, y)\phi(y)\, \mathrm{d}\sigma(y)\, \mathrm{d}s,$$

where we have used the Markov property of X. Now let $(t_k)_{k \in \mathbb{N}}$ denote a positive sequence which monotonically decreases to zero as $k \to \infty$. By the computation from above, we have for every $x \in \overline{D}$

$$\mathbb{E}_x \int_0^t \phi(X_s)\, \mathrm{d}L_s = \int_{t_k}^t \int_{\partial D} p(s,x,y)\phi(y)\, \mathrm{d}\sigma(y)\, \mathrm{d}s + \mathbb{E}_x \int_0^{t_k} \phi(X_s)\, \mathrm{d}L_s.$$

The claim follows by the facts that ϕ is bounded and $\mathbb{E}_x L_{t_k}$ goes to zero as $k \to \infty$ which follows from monotonicity and continuity of the local time and the property $L_0 = 0$ $\mathbb{P}_x$-a.s. for every $x \in \overline{D}$. $\qquad\square$

2.3.1 Continuum model

The main result for the continuum model (1.3), (1.5) is the following theorem.

Theorem 2.12. *Let f be a bounded Borel function satisfying $\langle f, 1 \rangle_{\partial D} = 0$. Then there is a unique weak solution $u \in C(\overline{D}) \cap H^1_\diamond(D)$ to the boundary value problem (1.3), (1.5). This solution admits the Feynman-Kac representation*

$$u(x) = \lim_{t \to \infty} \mathbb{E}_x \int_0^t f(X_s) \, dL_s \quad \text{for all } x \in \overline{D}. \tag{2.22}$$

Proof. The existence of a unique normalized weak solution u to (1.3), (1.5) is guaranteed by the standard theory of linear elliptic boundary value problems. Let us set

$$u_t(x) := \mathbb{E}_x \int_0^t f(X_s) \, dL_s \quad \text{and} \quad u_\infty(x) := \lim_{t \to \infty} u_t(x), \quad x \in \overline{D},$$

respectively. From the occupation formula (2.21) and the compatibility condition (1.6), it follows immediately that

$$u_t(x) = \int_0^t \int_{\partial D} (p(s, x, y) - |D|^{-1}) f(y) \, d\sigma(y) \, ds \quad \text{for all } x \in \overline{D}.$$

By Lemma 2.9 the convergence towards the stationary distribution is uniform over $\overline{D}$, in particular,

$$u_\infty(x) = \int_0^\infty \int_{\partial D} (p(s, x, y) - |D|^{-1}) f(y) \, d\sigma(y) \, ds \quad \text{for all } x \in \overline{D}. \tag{2.23}$$

It follows from (2.23) together with the Hölder continuity shown in Proposition 2.4 and the Aronson type upper bound (2.11) that u_∞ is in $C(\overline{D})$. Moreover, by the facts that p is the transition kernel density of a reflecting diffusion process and f is bounded, Fubini's theorem yields

$$\int_D \left(\int_0^\infty \int_{\partial D} p(s, x, y) f(y) \, d\sigma(y) \, ds \right) dx = \int_0^\infty \int_{\partial D} f(y) \, d\sigma(y) \, ds = 0,$$

in particular we obtain together with the regularity implied by the representation (2.23) that u_∞ has vanishing mean over D.

Now, let us use the following regularization technique in order to show $u \equiv u_\infty$: Let $(\kappa^{(k)})_{k \in \mathbb{N}}$ denote a sequence of smooth conductivities with components in $C^\infty(\overline{D})$ such that for $1 \leq i, j \leq d$, $\kappa_{ij}^{(k)} \to \kappa_{ij}$ a.e. as $k \to \infty$. Let us consider the Dirichlet forms $(\mathcal{E}^{(k)}, H^1(D))$, $k \in \mathbb{N}$, with

$$\mathcal{E}^{(k)}(v, w) := \int_D \kappa^{(k)} \nabla v \cdot \nabla w \, dx$$

and the associated reflecting diffusion processes $X^{(k)}$. By Proposition 2.5, we obtain the Skorohod decomposition

$$X_t^{(k)} = x + \int_0^t a^{(k)}(X_s^{(k)})\,\mathrm{d}s + \int_0^t B^{(k)}(X_s)\,\mathrm{d}W_s - \int_0^t \kappa^{(k)}(X_s^{(k)})\nu(X_s^{(k)})\,\mathrm{d}L_s^{(k)},$$

where W is a standard d-dimensional Brownian motion, $a_i^{(k)} := \sum_{j=1}^d \partial_j \kappa_{ij}^{(k)}$, $i = 1, ..., d$, and the matrix $B^{(k)}$ satisfies $2\kappa^{(k)} = (B^{(k)})^2$. Let us define $u_t^{(k)}$ in the same manner as u_t and $u^{(k)}(x) := \lim_{t \to \infty} u_t^{(k)}(x)$, $x \in \overline{D}$. We show that $u^{(k)}$ is the unique weak solution of the elliptic boundary value problem

$$\begin{cases} \nabla \cdot (\kappa^{(k)} \nabla u^{(k)}) = 0 & \text{in } D \\ \partial_{\kappa^{(k)}\nu} u^{(k)} = f & \text{on } \partial D \end{cases}$$

in the Sobolev space $H_\diamond^1(D)$. For test functions $v \in V_{\kappa^{(k)}}(D)$, we may apply Itô's formula for semimartingales to obtain

$$\mathbb{E}_x v(X_t^{(k)}) = v(x) + \mathbb{E}_x \int_0^t \nabla \cdot (\kappa^{(k)} \nabla v(X_s^{(k)}))\,\mathrm{d}s.$$

By Fubini's theorem, this is equivalent to

$$T_t^{(k)} v(x) - v(x) = \int_0^t \int_D p^{(k)}(s, x, y) \nabla \cdot (\kappa^{(k)} \nabla v(y))\,\mathrm{d}y\,\mathrm{d}s,$$

where we have used the superscript "(k)" for the semigroup and transition kernel density, respectively, corresponding to $\kappa^{(k)}$. Multiplication with f, integration over ∂D and another change of the orders of integration finally yield

$$\int_{\partial D} f(y)(T_t^{(k)} v(y) - v(y))\,\mathrm{d}\sigma(y) = \left\langle u_t^{(k)}, \nabla \cdot (\kappa^{(k)} \nabla v) \right\rangle.$$

Since $u_t^{(k)} \to u^{(k)}$ and $T_t^{(k)} v \to |D|^{-1} \int_D v\,\mathrm{d}x$, both uniformly on $\overline{D}$, as $t \to \infty$, we have

$$\left\langle u^{(k)}, \nabla \cdot (\kappa^{(k)} \nabla v) \right\rangle = - \left\langle f, v \right\rangle_{\partial D},$$

where we have used the expression (2.23) with $p^{(k)}$ instead of p for $u^{(k)}$. As this holds true for every $v \in V_{\kappa^{(k)}}(D)$, $u^{(k)}$ must be the unique normalized weak solution to the boundary value problem by a density argument.

Next, we show the convergence of the sequence $(u^{(k)})_{k\in\mathbb{N}}$ as $k \to \infty$ towards $u \in H^1_\diamond(D)$, the unique solution to (1.3), (1.5). From our assumptions on the sequence $(\kappa^{(k)})_{k\in\mathbb{N}}$, it is clear that $\kappa^{(k)}_{ij} \to \kappa_{ij}$, $1 \le i, j \le d$, in $L^2(D)$ as $k \to 0$, which implies G-convergence of the sequence of elliptic operators $(\mathcal{L}^{(k)})_{k\in\mathbb{N}}$ on D towards $\mathcal{L}$, cf. [165]. That is, for any $\phi \in (H^1_0(D))'$, the solutions $w^{(k)} \in H^1_0(D)$ to the homogeneous Dirichlet problem

$$\nabla \cdot (\kappa^{(k)} \nabla w^{(k)}) = \phi \quad \text{in } D \tag{2.24}$$

satisfy $w^{(k)} \rightharpoonup w$ in $H^1_0(D)$ as $k \to \infty$ and $\kappa^{(k)}\nabla w^{(k)} \rightharpoonup \kappa\nabla w$ in $L^2(D;\mathbb{R}^d)$ as $k \to \infty$, where $w \in H^1_0(D)$ is the solution to the homogeneous Dirichlet problem

$$\nabla \cdot (\kappa\nabla w) = \phi \quad \text{in } D.$$

Consider the variational form of the Neumann problem for $u^{(k)} \in H^1_\diamond(D)$, i.e.,

$$\int_D \kappa^{(k)}\nabla u^{(k)} \cdot \nabla v \, dx = \langle f, v \rangle_{\partial D} \quad \text{for all } v \in H^1_\diamond(D).$$

As $u^{(k)} \in H^1_\diamond(D)$ for all $k \in \mathbb{N}$, we have by the Poincaré inequality

$$\|u^{(k)}\| \le c_1\|\nabla u^{(k)}\|_2 \le c_2\|f\|_2.$$

That is, the sequence $(u^{(k)})_{k\in\mathbb{N}}$ is bounded in $H^1_\diamond(D)$ and we may extract a weakly convergent subsequence, still denoted $(u^{(k)})_{k\in\mathbb{N}}$ for convenience. We have $u^{(k)} \rightharpoonup \tilde{u}$ in $H^1_\diamond(D)$, as $k \to \infty$, so that it remains to show the convergence of the flows

$$\kappa^{(k)}\nabla u^{(k)} \rightharpoonup \kappa\nabla\tilde{u} \quad \text{in } L^2(D), \text{ as } k \to \infty.$$

We choose an arbitrary $w \in C^\infty_0(D)$, set $\phi := \nabla \cdot (\kappa\nabla w)$ and consider the solutions $w^{(k)}$, $k \in \mathbb{N}$, of the corresponding homogeneous Dirichlet problem (2.24). Then we have for every $k \in \mathbb{N}$ the trivial identity

$$\int_D \kappa^{(k)}\nabla w^{(k)} \cdot \nabla u^{(k)} \, dx = \int_D \kappa^{(k)}\nabla u^{(k)} \cdot \nabla w^{(k)} \, dx$$

and the convergence of the flows follows from the compensated compactness lemma [165, Lemma 1.1]. We may thus pass to the limit in the variational formulation to see that $\tilde{u} \equiv u$ is the unique solution to the Neumann problem.

On the other hand, by [146, Lemma 2.2] together with the Hölder continuity up to the boundary of both $p^{(k)}$, $k \in \mathbb{N}$, and p, it follows that for fixed $x \in \overline{D}$, $p^{(k)}(\cdot, x, \cdot) \to p(\cdot, x, \cdot)$ uniformly on compacts in $(0, T] \times \overline{D}$ for all $T > 0$. It follows from (2.23) that $u^{(k)}(x) \to u_\infty(x)$ for all $x \in \overline{D}$ as $k \to \infty$. Hence, u must coincide with u_∞ and the assertion is proved. $\square$

Remark 2.13. Note that the regularization technique we employed in the proof of Theorem 2.12 may be easily modified to prove the Feynman-Kac formula

$$u(x) = \mathbb{E}_x \phi(X_{\tau(D)}), \quad x \in D$$

for the conductivity equation (1.3) with Dirichlet boundary condition $u|_{\partial D} = \phi$, where $\phi \in H^{1/2}(D)$ and

$$\tau(D) := \inf\{t \geq 0 : X_t \in \mathbb{R}^d \backslash D\}$$

denotes the *first exit time* from the domain D. Such a proof requires the fact that

$$X^{(k)}_{\cdot \wedge \tau^{(k)}(D)} \to X_{\cdot \wedge \tau(D)} \quad \text{in law on } C([0, \infty); \overline{D})$$

as $n \to \infty$ for every $x \in D$. This follows immediately from the Lipschitz property of ∂D, implying that all points of ∂D are *regular* in the sense of [80, Chapter 4.2], see also the proof of Theorem 3.2 in the next chapter.

A slight modification of the arguments from above yields the following result which is in fact a corollary rather to the proof of Theorem 2.12 than to its actual statement.

Corollary 2.14. *Let f be a bounded Borel function and let $\alpha > 0$. Then there is a unique weak solution $u \in C(\overline{D}) \cap H^1_\diamond(D)$ to the boundary value problem*

$$\nabla \cdot (\kappa \nabla u) - \alpha u = 0 \quad \text{in } D$$

subject to the boundary condition (1.5). This solution admits the Feynman-Kac representation

$$u(x) = \mathbb{E}_x \int_0^\infty e^{-\alpha t} f(X_t) \, \mathrm{d}L_t \quad \text{for all } x \in \overline{D}. \tag{2.25}$$

Proof. Repeat the proof of Theorem 2.12, however, substituting $\{T_t, t \geq 0\}$ with the *Feynman-Kac semigroup* $\{\widetilde{T}_t, t \geq 0\}$, $\widetilde{T}_t v(x) := \mathbb{E}_x e^{-\alpha t} v(X_t)$. Note that in contrast to the Neumann problem without the zero-order term, the *gauge function* $\mathbb{E}_x \int_0^\infty e^{-t} \, \mathrm{d}L_t$ is finite $\mathbb{P}_x$-a.s. for every $x \in \overline{D}$. $\square$

2.3.2 Complete electrode model

The main result for the complete electrode model (1.3), (1.12) is the following theorem.

Theorem 2.15. *For given functions f, g defined by (1.13) and a voltage pattern $U \in \mathbb{R}^N$ satisfying (1.10), there is a unique weak solution $u \in C(\overline{D}) \cap H^1(D)$ to the boundary value problem (1.3), (1.12). This solution admits the Feynman-Kac representation*

$$u(x) = \mathbb{E}_x \int_0^\infty e_g(t) f(X_t) \, dL_t \quad \text{for all } x \in \overline{D}, \tag{2.26}$$

with

$$e_g(t) := \exp\Big(-\int_0^t g(X_s) \, dL_s \Big), \quad t \geq 0. \tag{2.27}$$

Before we are ready to give a proof of Theorem 2.15, let us introduce the *Feynman-Kac semigroup* of the complete electrode model, i.e., the one-parameter family of operators $\{T_t^g, t \geq 0\}$ defined by

$$T_t^g v(x) := \mathbb{E}_x e_g(t) v(X_t), \quad x \in \overline{D} \text{ and } t \geq 0. \tag{2.28}$$

Let us define the *perturbed* Dirichlet form $(\mathcal{E}^g, \mathcal{D}(\mathcal{E}^g))$ by a perturbation of $(\mathcal{E}, \mathcal{D}(\mathcal{E}))$ with the measure $g \cdot \sigma$, i.e.,

$$\mathcal{E}^g(v, w) = \mathcal{E}(v, w) + \langle gv, w \rangle_{\partial D}, \quad v, w \in \mathcal{D}(\mathcal{E}^g), \tag{2.29}$$

where $\mathcal{D}(\mathcal{E}^g) = H^1(D)$ by the standard trace theorem.

Lemma 2.16. *The semigroup $\{T_t^g, t \geq 0\}$ is associated with the Dirichlet form $(\mathcal{E}^g, \mathcal{D}(\mathcal{E}^g))$.*

Proof. Let $G_\alpha^g \phi$, $\alpha > 0$, denote the Laplace transform of T_t^g

$$G_\alpha^g \phi(x) = \mathbb{E}_x \int_0^\infty e_g(t) e^{-\alpha t} \phi(X_t) \, dt.$$

By the one-to-one correspondence in Theorem A.6 it is sufficient to show that

$$G_\alpha^g \phi \in H^1(D) \text{ and } \mathcal{E}_\alpha^g(G_\alpha^g \phi, v) = \langle \phi, v \rangle \text{ for all } \phi \in L^2(D), \ v \in H^1(D).$$

We may assume without loss of generality that ϕ is a bounded Borel function: Due to the contraction property of the resolvent, we have

$$\|G_\alpha^g \phi\|_2 \leq \|G_\alpha \phi\|_2 \leq \alpha^{-1} \|\phi\|_2.$$

Suppose the claim holds for bounded Borel functions in $L^2(D)$ and that ϕ is non-negative, but not necessarily bounded. Choose $\phi_k := \phi \wedge k$, $k \in \mathbb{N}$, so that by the continuity of $G_\alpha g$ in $L^2(D)$ we can extend the claim from ϕ_k to ϕ. Finally, the claim for general ϕ follows by linearity.

For every $x \in \overline{D}$, we can write the expression $G_\alpha\phi(x) - G_\alpha^g\phi(x)$ equivalently as

$$
\mathbb{E}_x \int_0^\infty e^{-\alpha t} e^{-\int_0^t g(X_s)\,\mathrm{d}L_s} \left(e^{\int_0^t g(X_s)\,\mathrm{d}L_s} - 1 \right) \phi(X_t)\,\mathrm{d}t
$$

$$
= \mathbb{E}_x \int_0^\infty \left(\int_0^t e^{\int_0^s g(X_r)\,\mathrm{d}L_r} g(X_s)\,\mathrm{d}L_s \right) e^{-\alpha t} e^{-\int_0^t g(X_s)\,\mathrm{d}L_s} \phi(X_t)\,\mathrm{d}t
$$

$$
= \mathbb{E}_x \int_0^\infty e^{\int_0^s g(X_r)\,\mathrm{d}L_r} g(X_s) \left(\int_s^\infty e^{-\alpha t} e^{-\int_0^t g(X_s)\,\mathrm{d}L_s} \phi(X_t)\,\mathrm{d}t \right) \mathrm{d}L_s
$$

$$
= \mathbb{E}_x \int_0^\infty e^{-\alpha s} g(X_s) \left(\int_0^\infty e^{-\alpha t} e^{-\int_0^t g(X_s)\,\mathrm{d}L_s \circ \Theta_s} \phi(X_t \circ \Theta_s)\,\mathrm{d}t \right) \mathrm{d}L_s
$$

$$
= \mathbb{E}_x \int_0^\infty e^{-\alpha s} g(X_s) \mathbb{E}_{X_s} \left\{ \int_0^\infty e^{-\alpha t} e^{-\int_0^t g(X_r)\,\mathrm{d}L_r} \phi(X_t)\,\mathrm{d}t \right\} \mathrm{d}L_s,
$$

where we have used Fubini's theorem, the Markov property of X and the identity

$$
\exp\left(\int_0^t g(X_s)\,\mathrm{d}L_s \right) - 1 = \int_0^t \exp\left(\int_0^s g(X_r)\,\mathrm{d}L_r \right) g(X_s)\,\mathrm{d}L_s.
$$

We have thus shown

$$
G_\alpha\phi - G_\alpha^g\phi = \mathbb{E}_x \int_0^\infty e^{-\alpha t} g(X_t) G_\alpha^g\phi(X_t)\,\mathrm{d}L_t \quad \text{for all } \alpha > 0. \tag{2.30}
$$

By Corollary 2.14, the right-hand side in (2.30) is the unique weak solution to the elliptic boundary value problem

$$
\begin{cases}
\nabla \cdot (\kappa \nabla w) - \alpha w = 0 & \text{in } D \\
\partial_{\kappa\nu} w = g G_\alpha^g\phi & \text{on } \partial D.
\end{cases}
$$

That is, it coincides with the α-potential $U_\alpha(G_\alpha^g\phi(g \cdot \sigma)) \in H^1(D)$ and from the one-to-one correspondence in Theorem A.6 we obtain

$$
\mathcal{E}_\alpha(G_\alpha^g\phi, v) = \mathcal{E}_\alpha(G_\alpha\phi, v) + \mathcal{E}_\alpha(U_\alpha(G_\alpha^g\phi(g \cdot \sigma)), v) = \langle \phi, v \rangle - \langle G_\alpha^g\phi, gv \rangle_{\partial D}.
$$

The assertion is proved. $\qquad\square$

Proposition 2.17. *The Feynman-Kac semigroup $\{T_t^g, t \geq 0\}$ is a strong Feller semigroup on $L^2(D)$.*

Proof. First, it follows from Lemma 2.16 that $\{T_t^g, t \geq 0\}$ is a strongly continuous sub-Markovian contraction semigroup on $L^2(D)$. T_t^g is a bounded operator from $L^1(D)$ to $L^\infty(D)$ for every $t > 0$, which can be shown using Fatou's lemma. By the Dunford-Pettis theorem, T^g can thus be represented as an integral operator for every $t > 0$,

$$T_t^g \phi(x) = \int_D p^g(t, x, y) \phi(y) \, \mathrm{d}y \quad \text{for every } \phi \in L^1(D), \tag{2.31}$$

where for all $t > 0$ we have $p^g(t, x, y) \in L^\infty(D \times D)$ and $p^g(t, x, y) \geq 0$ for a.e. $x, y \in \overline{D}$. In order to prove the strong Feller property, we show that T_t^g, $t > 0$, maps bounded Borel functions to $C(\overline{D})$. As in the papers [70, 133], we use an iterative method to construct the transition kernel density p^g. Let $p_0^g(t, x, y) := p(t, x, y)$ and set

$$p_k^g(t, x, y) := -\int_0^t \int_{\partial D} p(s, x, z) g(z) p_{k-1}^g(t - s, z, y) \, \mathrm{d}\sigma(z) \, \mathrm{d}s, \quad k \in \mathbb{N}.$$

Note that the terms p_k^g are symmetric in the x and y variables by the symmetry of p. By induction, using Lemma 2.11, it is not difficult to verify that for all $k \in \mathbb{N}$

$$\int_0^t \int_{\partial D} g(x) p_k^g(s, x, y) \, \mathrm{d}\sigma(x) \, \mathrm{d}s \leq \left(\sup_{x \in \overline{D}} \left\{ \mathbb{E}_x \int_0^t g(X_s) \, \mathrm{d}L_s \right\} \right)^{k+1}$$

and that there is a positive constant c_1 such that

$$|p_k^g(t, x, y)| \leq c_1^{k+1} t^{-d/2} \left(\sup_{x \in \overline{D}} \left\{ \mathbb{E}_x \int_0^t g(X_s) \, \mathrm{d}L_s \right\} \right)^{k+1} \quad \text{for all } k \in \mathbb{N}. \tag{2.32}$$

Let us show the continuity of p_k^g, $k \in \mathbb{N} \cup \{0\}$. For $k = 0$, this is a consequence of Proposition 2.4. Now assume that p_{k-1}^g is continuous on $(t_0, T] \times \overline{D} \times \overline{D}$ for $t_0 > 0$, then we have for $t \in (t_0, T]$

$$p_k^g(t, x, y) = -\int_0^{t_0} \int_{\partial D} p(s, x, z) g(z) p_{k-1}^g(t - s, z, y) \, \mathrm{d}\sigma(z) \, \mathrm{d}s$$

$$- \int_{t_0}^t \int_{\partial D} p(s, x, z) g(z) p_{k-1}^g(t - s, z, y) \, \mathrm{d}\sigma(z) \, \mathrm{d}s.$$

Note that the first integral on the right-hand side tends to zero uniformly as $t_0 \to 0$, which is a consequence of (2.32), while the second integral is continuous by assumption. Hence, there exists a $T > 0$ such that the series $p^g(t, x, y) := \sum_{k=0}^{\infty} p_k^g(t, x, y)$ converges absolutely and uniformly on any compact subset of $(0, T] \times \overline{D} \times \overline{D}$ and is thus continuous on $(0, T] \times \overline{D} \times \overline{D}$. By the Markov property we have for all $t \in (0, T]$ and every $x \in \overline{D}$ the following expression for $T_t^g \phi(x)$

$$\int_D p^g(t, x, y)\phi(y)\, \mathrm{d}y = \mathbb{E}_x \phi(X_t) + \sum_{k=1}^{\infty} \frac{1}{k!} \mathbb{E}_x \left\{ \left(- \int_0^t g(X_s)\, \mathrm{d}L_s \right)^k \phi(X_t) \right\}.$$

Therefore, the assertion for arbitrary $T > 0$ follows from the Chapman-Kolmogorov equation. $\qquad\square$

Corollary 2.18. *Let u be defined by the Feynman-Kac formula (2.26), then $u \in C(\overline{D})$.*

Proof. Let us define a $\mathbb{P}_x$-martingale by

$$\mathbb{E}_x \left\{ \int_0^{\infty} e_g(s)f(X_s)\, \mathrm{d}L_s \Big| \mathcal{F}_t \right\} = \int_0^t e_g(s)f(X_s)\, \mathrm{d}L_s + e_g(t)u(X_t),$$

where the right-hand side is obtained using the Markov property of X together with the fact that e_g is a multiplicative functional of X. Obviously,

$$e_g(t)u(X_t) - u(x) + \int_0^t e_g(s)f(X_s)\, \mathrm{d}L_s$$

is a $\mathbb{P}_x$-martingale, as well, and hence we have for all $0 \leq s \leq t$

$$e_g(s)u(X_s) = e_g(s)\mathbb{E}_{X_s} e_g(t-s)u(X_{t-s}) + e_g(s)\mathbb{E}_{X_s} \int_0^{t-s} e_g(r)f(X_r)\, \mathrm{d}L_r.$$

Setting $s = 0$ thus yields

$$u(x) = T_t^g u(x) + \mathbb{E}_x \int_0^t e_g(s)f(X_s)\, \mathrm{d}L_s \quad \text{for all } t \geq 0. \tag{2.33}$$

$T_t^g u$ is continuous on $\overline{D}$ by Proposition 2.17. To prove that u is continuous on $\overline{D}$, it is sufficient to show that the second term on the right-hand side of (2.33) tends to zero uniformly in x, as $t \to 0$. This is, however, clear since we may estimate

$$\sup_{x \in \overline{D}} \left\{ \mathbb{E}_x \int_0^t e_g(s)f(X_s)\, \mathrm{d}L_s \right\} \leq z^{-1} \max_{l=1,\ldots,N} \{U_l\} \sup_{x \in \overline{D}} \{\mathbb{E}_x L_t\},$$

where the right-hand side tends to zero as $t \to 0$ by Lemma 2.11. $\qquad\square$

The following lemma yields a semimartingale decomposition for the composite process $u(X_t)$ which compensates for the lack of Itô's formula in the proof of Theorem 2.15.

Lemma 2.19. *Let $u \in H^1(D)$ denote the weak solution to the boundary value problem (1.3), (1.12). Then for all $t \geq 0$*

$$u(X_t) = u(x) + \int_0^t \nabla u(X_s)\, \mathrm{d}M_s^u - \int_0^t f(X_s)\, \mathrm{d}L_s + \int_0^t g(X_s)u(X_s)\, \mathrm{d}L_s,$$
(2.34)

$\mathbb{P}_x$-*a.s. for q.e. $x \in \overline{D}$.*

Proof. Applying the Fukushima decomposition (2.13) to the perturbed Dirichlet form $(\mathcal{E}^g, H^1(D))$, we obtain the unique decomposition

$$v(X_t^g) - v(X_0^g) = M_t^{g,v} + N_t^{g,v}, \quad \text{for all } t > 0, \quad \mathbb{P}_x\text{-a.s. for q.e. } x \in \overline{D}$$

into a martingale additive functional of finite energy and a continuous additive functional of zero energy of the non-conservative Hunt process X^g associated with $(\mathcal{E}^g, H^1(D))$. We study the relation between the continuous additive functionals N^v and $N^{g,v}$. Let us assume first that v is in the range of the 1-resolvent associated with the perturbed Dirichlet form, i.e.,

$$v(x) = G_1^g \phi(x) = \mathbb{E}_x \int_0^\infty e^{-t - \int_0^t g(X_s)\, \mathrm{d}L_s} \phi(X_t)\, \mathrm{d}t$$

for some $\phi \in L^2(D)$. Then we have the identity $-\mathcal{L}G_1^g\phi = \phi - v$ so that for all $w \in H^1(D)$

$$\mathcal{E}^g(G_1^g\phi, w) = \int_D (\phi - v)w\, \mathrm{d}x.$$

Using Lemma A.29 and the Revuz correspondence, we see that $N^{g,v}$ admits a semimartingale decomposition, namely

$$N_t^{g,v} = \int_0^t (\phi(X_s^g) - v(X_s^g))\, \mathrm{d}s.$$

Moreover, we have shown in the proof of Lemma 2.16 that

$$G_1\phi - G_1^g\phi = U_1(vg \cdot \sigma).$$

Invoking Lemma A.29 in conjunction with the Revuz correspondence, we see that the zero energy continuous additive functional in the Fukushima decomposition of the 1-potential corresponding to the signed Radon measure $vg \cdot \sigma$ is given by

$$N_t^{U_1(vg \cdot \sigma)} = \int_0^t U_1(vg \cdot \sigma)(X_s) \, ds - \int_0^t v(X_s) g(X_s) \, dL_s \quad \text{for all } t \geq 0.$$

Therefore, we obtain

$$
\begin{aligned}
N_t^v &= N_t^{G_1 \phi} - N_t^{U_1(vg \cdot \sigma)} \\
&= \int_0^t (\phi(X_s) - G_1 \phi(X_s)) \, ds - \int_0^t U_1(vg \cdot \sigma)(X_s) \, ds \\
&\quad + \int_0^t v(X_s) g(X_s) \, dL_s \\
&= \int_0^t (\phi(X_s) - v(X_s)) \, ds + \int_0^t v(X_s) g(X_s) \, dL_s.
\end{aligned}
$$

Moreover, notice that X^g is related to X by a random time change, namely

$$X_s^g = \begin{cases} X_s, & s < \zeta^g \\ \partial, & s \geq \zeta^g, \end{cases}$$

where the lifetime ζ^g is given by

$$\zeta^g := \inf \left\{ t : \int_0^t g(X_s) \, dL_s > Z \right\}$$

and Z is an exponentially distributed random variable with parameter 1. Hence, we obtain

$$N_t^v = N_t^{g,v} + \int_0^t v(X_s^g) g(X_s^g) \, dL_s \quad \text{for all } t < \zeta^g. \tag{2.35}$$

This equality may be generalized to the case of an arbitrary $v \in H^1(D)$ not necessarily in the range of the resolvent using an approximation argument. Namely, we consider the sequence $(v_k)_{k \in \mathbb{N}}$ with $v_k := k G_{k+1}^g v = G_1^g \phi_k$, $\phi_k := k(v - k G_{k+1}^g v)$. Then $v_k \in H^1(D)$ for all $k \in \mathbb{N}$ and the sequence $(v_k)_{k \in \mathbb{N}}$ satisfies both

$$\lim_{k \to \infty} \mathcal{E}^g(v_k - v, v_k - v) = 0 \quad \text{and} \quad \lim_{k \to \infty} \mathcal{E}(v_k - v, v_k - v) = 0$$

so that by [54, Corollary 5.2.1], there exists a subsequence, for convenience still denoted $(v_k)_{k \in \mathbb{N}}$, such that we have $v_k(X_t^g) \to v(X_t^g)$, $N_t^{g,v_k} \to N_t^{g,v}$

and $N_t^{v_k} \to N_t^v$ uniformly on each finite time interval, $\mathbb{P}_x$-a.s. for q.e. $x \in \overline{D}$. In particular, it follows that (2.35) holds for arbitrary $v \in H^1(D)$.

As u solves the boundary value problem (1.3), (1.12) we have that

$$\mathcal{E}^g(u,v) = \langle f, v \rangle_{\partial D} \quad \text{for all } v \in H^1(D) \cap C(\overline{D}).$$

Since the perturbed Dirichlet form $(\mathcal{E}^g, H^1(D))$ is regular, we may apply Lemma A.29 with $\mathcal{E}^g$ instead of $\mathcal{E}$, which yields together with the Revuz correspondence the representation

$$N_t^{g,u} = -\int_0^t f(X_s^g)\,\mathrm{d}L_s,$$

$\mathbb{P}_x$-a.s. for q.e. $x \in \overline{D}$. Finally, using the representation

$$M_t^u = \int_0^t \nabla u(X_s)\,\mathrm{d}M_s^u,$$

$\mathbb{P}_x$-a.s. for q.e. $x \in \overline{D}$ for the martingale additive functional, the claim follows from the Fukushima decomposition (2.13), (2.35) and the Markov property of X. $\qquad\square$

Proof of Theorem 2.15. There exists a weak solution $u \in H^1(D)$ of the boundary value problem (1.3), (1.12) so that with regard to Corollary 2.18, it remains to show that this weak solution u admits the Feynman-Kac representation (2.26). Note first that the gauge function

$$\mathbb{E}_x \int_0^\infty e_g(t)\,\mathrm{d}L_t \tag{2.36}$$

is finite $\mathbb{P}_x$-a.s. for every $x \in \overline{D}$, hence the expression on the right-hand side of (2.26) is well-defined. Lemma 2.19 yields the semimartingale decomposition

$$u(X_t) = u(x) + \int_0^t \nabla u(X_s)\,\mathrm{d}M_s^u - \int_0^t f(X_s)\,\mathrm{d}L_s + \int_0^t g(X_s)u(X_s)\,\mathrm{d}L_s,$$

$\mathbb{P}_x$-a.s. for q.e. $x \in \overline{D}$. Note that the second term on the right-hand side is a local $\mathbb{P}_x$-martingale and that e_g is continuous, adapted to $\{\mathcal{F}_t, t \geq 0\}$ and of bounded variation. Multiplication by such functions leaves the class of semimartingales invariant. Using integration by parts we thus obtain for q.e. $x \in \overline{D}$ the identity

$$u(X_t)e_g(t) = u(x) + \int_0^t e_g(s)\nabla u(X_s)\,\mathrm{d}M_s^u - \int_0^t e_g(s)f(X_s)\,\mathrm{d}L_s,$$

$\mathbb{P}_x$-a.s., where the second summand on the right-hand side is a local $\mathbb{P}_x$-martingale. That is, there exists an increasing sequence $(\tau_k)_{k \in \mathbb{N}}$ of stopping times which tend to infinity such that for every $k \in \mathbb{N}$

$$\mathcal{M}_{t \wedge \tau_k} := \int_0^{t \wedge \tau_k} e_g(s) \nabla u(X_s) \, \mathrm{d}M_s^u$$

is a $\mathbb{P}_x$-martingale. By definition of the term e_g, it is, however, clear that

$$\mathbb{E}_x \sup_{k \in \mathbb{N}} |\mathcal{M}_{t \wedge \tau_k}| < \infty \quad \text{for all } t \geq 0 \text{ and every } x \in \overline{D}$$

which is sufficient for $\{\mathcal{M}_t, t \geq 0\}$ to be a $\mathbb{P}_x$-martingale by the dominated convergence theorem. Hence,

$$u(x) = \mathbb{E}_x \int_0^t e_g(s) f(X_s) \, \mathrm{d}L_s + \mathbb{E}_x u(X_t) e_g(t) \quad \text{for q.e. } x \in \overline{D}.$$

Letting $t \to \infty$ finally yields

$$u(x) = \mathbb{E}_x \int_0^\infty e_g(t) f(X_t) \, \mathrm{d}L_t \quad \text{for q.e. } x \in \overline{D},$$

where we have used the fact that u is essentially bounded by standard elliptic regularity theory. Finally, by the fact that we have actually shown in Corollary 2.18 that the right-hand side in the last equality is continuous up to the boundary, the assertion holds for every $x \in \overline{D}$. $\qquad \square$

Remark 2.20. Note that the technique we used to prove Theorem 2.15 fails for the Neumann problem corresponding to the continuum model. This comes from the fact that in this case the gauge function (2.36) becomes infinite. For the same reason Theorem 1.2 from [34], specialized to a zero lower-order term, does not yield the desired Feynman-Kac formula for the continuum model either.

2.3.3 Anomaly detection mixed boundary value problem

Now we can directly deduce the desired Feynman-Kac formula for the mixed boundary value problem corresponding to the stochastic anomaly detection problem introduced in Section 1.3 of the previous chapter. Recall that in this setting ∂D consists of two disjoint parts $\partial_1 D$ and $\partial_2 D$ and that measurements can be taken only on the *accessible boundary* $\partial_1 D$ while the electric potential vanishes on the *inaccessible boundary* $\partial_2 D$. The deterministic EIT forward problem for the complete electrode model is then given by the conductivity equation (1.3) subject to the mixed boundary conditions

$$\kappa\nu \cdot \nabla u|_{\partial D} + gu|_{\partial D} = f \qquad \text{on } \partial_1 D$$
$$u|_{\partial D} = 0 \qquad \text{on } \partial_2 D. \tag{2.37}$$

The following result is a corollary to the line of arguments that led to the proof of Theorem 2.15 rather than to its actual statement.

Corollary 2.21. *For given functions f, g defined by (1.13) and a voltage pattern $U \in \mathbb{R}^N$ satisfying (1.10), there is a unique weak solution $u \in C(\overline{D}) \cap H^1(D)$ to the boundary value problem (1.3), (2.37). This solution admits the Feynman-Kac representation*

$$u(x) = \mathbb{E}_x \int_0^{\tau} e_g(t) f(X_t) \, \mathrm{d}L_t \quad \text{for all } x \in \overline{D}, \tag{2.38}$$

where $\tau := \inf\{t \geq 0 : X_t \in \partial_2 D\}$ denotes the first hitting time of $\partial_2 D$.

Proof. Repeat the arguments from Subsection 2.3.2 with the Feynman-Kac semigroup $\{\widetilde{T}_t^g, t \geq 0\}$, where $\widetilde{T}_t^g v(x) := \mathbb{E}_x\{[t \leq \tau] e_g(t) v(X_t)\}$ instead of $\{T_t^g, t \geq 0\}$. $\qquad\square$

Part II

Anomaly detection in heterogeneous media

3 Stochastic homogenization: Theory and numerics

In this chapter, we pursue two related goals. First, we derive a theoretical *stochastic* homogenization result for the stochastic forward problem introduced in the first chapter. The key ingredient to obtain this result is the use of the Feynman-Kac formula for the complete electrode model. The proof is constructive in the sense that it yields a strategy to achieve our second goal, the numerical approximation of the *effective* conductivity. In contrast to periodic homogenization, which is well understood, numerical homogenization of random media still poses major practical challenges. In order to cope with these challenges, we propose a new numerical method inspired by a highly efficient stochastic method from the physics literature that was invented by Torquato, Kim and Cule [159]. From a mathematical point of view, the novelty of the proposed method lies in the fact that it is based on the aforementioned stochastic homogenization result, that is, on simulation of the underlying diffusion process rather than on the usual discretization of the so-called *auxiliary problem*.

We start in Section 3.1 by recalling some preliminaries from homogenization theory. Then, in Section 3.2, we use the Feynman-Kac formula obtained in the previous chapter to establish a rigorous link between homogenization of the stochastic forward problem for the complete electrode model and homogenization of the underlying diffusion process. In Section 3.3, we introduce a numerical method based on this result; in order to emphasize the difference to homogenization via discrete lattice random walks, we call this method a *continuum micro-scale Monte Carlo method*. We discuss path simulation in digitized random media and provide a convergence analysis for the proposed method. Finally, in Section 3.4, we present numerical examples to support our findings.

3.1 Preliminaries

For convenience of the reader, let us recall some standard concepts from homogenization theory. Let $\phi := (\phi_1, ..., \phi_d)$, $\phi_i \in L^2_{\text{loc}}(\mathbb{R}^d)$, $i = 1, ..., d$, denote a vector field. We say that ϕ is a *gradient field* if for every $\psi \in C^\infty_c(\mathbb{R}^d)$,

$$\int_{\mathbb{R}^d} \phi_i \partial_j \psi - \phi_j \partial_i \psi \, \mathrm{d}x = 0, \quad i, j = 1, ..., d.$$

Moreover, we say that ϕ is *divergence-free* if for every $\psi \in C^\infty_c(\mathbb{R}^d)$,

$$\sum_{i=1}^d \int_{\mathbb{R}^d} \phi_i \partial_i \psi \, \mathrm{d}x = 0.$$

Now let us consider a conductivity random field $\{\kappa(x, \omega), (x, \omega) \in \mathbb{R}^d \times \Gamma\}$ and let $\{\Theta_x, x \in \mathbb{R}^d\}$ denote the underlying dynamical system which is assumed to satisfy the assumptions (i)-(iv) from Subsection 1.3.1. A vector field $\phi \in L^2(\Gamma; \mathbb{R}^d)$ is called a *gradient field*, respectively *divergence-free*, if its realizations $\phi(\cdot, \omega) : \mathbb{R}^d \mapsto \mathbb{R}^d$, $x \mapsto \phi(\Theta_x \omega)$ are gradient fields, respectively divergence-free, for $\mathcal{P}$-a.e. $\omega \in \Gamma$. We define the function spaces

$$L^2_{\text{pot}}(\Gamma) := \{\phi \in L^2(\Gamma; \mathbb{R}^d) : \phi(\cdot, \omega) \text{ is a gradient field } \mathcal{P}\text{-a.s.}\}$$

$$L^2_{\text{sol}}(\Gamma) := \{\phi \in L^2(\Gamma; \mathbb{R}^d) : \phi(\cdot, \omega) \text{ is divergence-free } \mathcal{P}\text{-a.s.}\}.$$

If $\phi \in L^2_{\text{pot}}(\Gamma)$, we can find a function $\eta : \mathbb{R}^d \times \Gamma \to \mathbb{R}$ such that $\eta(\cdot, \omega) \in H^1_{\text{loc}}(\mathbb{R}^d)$ and

$$\nabla \eta(\cdot, \omega) = \phi(\Theta.\omega) \quad \text{a.e. in } \mathbb{R}^d \text{ for } \mathcal{P}\text{-a.e. } \omega \in \Gamma. \tag{3.1}$$

In particular, (3.1) defines a stationary random field with respect to the measure $\mathcal{P}$. We call η the *potential* corresponding to ϕ.

Remark 3.1. Note that $\phi \in L^2_{\text{pot}}(\Gamma)$ does not imply that $\{\eta(x, \omega), (x, \omega) \in \mathbb{R}^d \times \Gamma\}$ is a stationary random field with respect to $\mathcal{P}$. In fact, it can be shown that this is not true for $d = 1$.

We define another function space

$$\mathcal{V}^2_{\text{pot}} := \{\phi \in L^2_{\text{pot}}(\Gamma) : M\phi = 0\},$$

so that one obtains an orthogonal Weyl decomposition of $L^2(\Gamma; \mathbb{R}^d)$, namely

$$L^2(\Gamma; \mathbb{R}^d) = \mathcal{V}^2_{\text{pot}}(\Gamma) \oplus L^2_{\text{sol}}(\Gamma),$$

cf., e.g., [165]. Let $\xi \in \mathbb{R}^d$ denote a direction vector, i.e., $|\xi| = 1$. The so-called *auxiliary problem* for the direction ξ reads as follows: Find $\chi^\xi \in \mathcal{V}^2_{\mathrm{pot}}(\Gamma)$ such that $\kappa(\xi + \chi^\xi) \in L^2_{\mathrm{sol}}(\Gamma)$ or equivalently,

$$\mathrm{M}\{\kappa(\xi + \chi^\xi) \cdot \phi\} = 0 \quad \text{for all } \phi \in \mathcal{V}^2_{\mathrm{pot}}(\Gamma). \tag{3.2}$$

For a proof of existence and uniqueness of the solution to the auxiliary problem we refer the reader to the seminal paper [132] by Papanicolaou and Varadhan.

We can now bring the underlying diffusion processes evolving in the random medium into play by recalling a stochastic homogenization result which was obtained by Lejay [104]. Let $\{\kappa_\varepsilon(x, \omega), (x, \omega) \in \mathbb{R}^d \times \Gamma\}$ denote the scaled conductivity random field, see Subsection 1.3.1, and let $X^{\omega,\varepsilon}$ denote the diffusion process on $\mathbb{R}^d$ which is associated with the regular symmetric Dirichlet form

$$\mathcal{E}^{\omega,\varepsilon}(v, w) := \int_{\mathbb{R}^d} \kappa_\varepsilon(\cdot, \omega) \nabla v \cdot \nabla w \, dx, \quad v, w \in \mathcal{D}(\mathcal{E}^{\omega,\varepsilon}) := H^1(\mathbb{R}^d)$$

on $L^2(\mathbb{R}^d)$. It has been shown in [104] that, under assumption (A1) from Subsection 1.3.1, for $\mathcal{P}$-a.e. $\omega \in \Gamma$

$$X^{\omega,\varepsilon} \to X^* \quad \text{in law on } C([0, \infty); \mathbb{R}^d) \text{ as } \varepsilon \to 0, \tag{3.3}$$

where X^* denotes the diffusion process on $\mathbb{R}^d$ which is associated with the *homogenized Dirichlet form*

$$\mathcal{E}^*(v, w) := \int_{\mathbb{R}^d} \kappa^* \nabla v \cdot \nabla w \, dx, \quad v, w \in \mathcal{D}(\mathcal{E}^*) := H^1(\mathbb{R}^d)$$

on $L^2(\mathbb{R}^d)$ and the constant, symmetric and positive definite matrix κ^* satisfies the equation

$$\xi \cdot \kappa^* \xi = 2\mathrm{M}\{(\xi + \chi^\xi) \cdot \kappa(\xi + \chi^\xi)\}, \tag{3.4}$$

where $\chi^\xi \in \mathcal{V}^2_{\mathrm{pot}}(\Gamma)$ denotes the solution to the auxiliary problem (3.2) for the direction $\xi \in \mathbb{R}^d$.

3.2 Homogenization of the stochastic forward problem

In this section, we derive a homogenization result for the stochastic forward problem introduced in the first chapter thus providing a rigorous link between homogenization of the boundary value problem and homogenization of the underlying diffusion process. We will deduce from this result that stochastic numerical homogenization methods can be obtained from discretizing diffusion processes on $\mathbb{R}^d$ which is indeed the theoretical foundation for the method proposed in Section 3.3 below.

The following theorem is our main result. Its assertion is in fact a rather direct consequence of an invariance principle for reflecting diffusion processes obtained recently by Chen, Croydon and Kumagai [33] and the Feynman-Kac formula (2.38) from Corollary 2.21.

Theorem 3.2. *Let* $D = B(0,R) \cap \mathbb{R}^d_-$, $R > 0$, *and let* $\{\kappa_\varepsilon(x,\omega), (x,\omega) \in \mathbb{R}^d \times \Gamma\}$ *be a stationary random field satisfying assumption (A1) from Subsection 1.3.1. Assume moreover that the trajectories of the random field satisfy* $\kappa(\cdot,\omega) \in C^{0,1}_{loc}(\overline{D}; \mathbb{R}^{d \times d})$ *or* $\kappa(\cdot,\omega)$ *piecewise constant for* $\mathcal{P}$*-a.e.* $\omega \in \Gamma$; *let* $\Sigma = \emptyset$. *Then, for a given voltage pattern* $U \in \mathbb{R}^N$, *we have for the potentials in the stochastic boundary value problem (1.16), (1.17)*

$$u_\varepsilon(x,\omega) \to u^*(x), \quad x \in \overline{D} \text{ for } \mathcal{P}\text{-a.e. } \omega \in \Gamma, \text{ as } \varepsilon \to 0, \qquad (3.5)$$

and the corresponding electrode currents satisfy

$$\lim_{\varepsilon \to 0} J_l(\varepsilon,\omega) = \frac{1}{|E_l|} \int_{E_l} \kappa^* \nu \cdot \nabla u^*(x)|_{\partial_1 D} \, d\sigma(x) \quad \text{for } \mathcal{P}\text{-a.e. } \omega \in \Gamma, \qquad (3.6)$$

$l = 1, ..., N$, *where the function* $u^* \in H^1_0(D \cup \partial_1 D) \cap C(\overline{D})$ *is the unique solution to the deterministic forward problem*

$$\nabla \cdot (\kappa^* \nabla u^*) = 0 \quad \text{in } D \qquad (3.7)$$

subject to the boundary conditions

$$\begin{aligned} \kappa^* \nu \cdot \nabla u^*|_{\partial_1 D} + g u^*|_{\partial_1 D} &= f \qquad \text{on } \partial_1 D \\ u^*|_{\partial_2 D} &= 0 \qquad \text{on } \partial_2 D \end{aligned} \qquad (3.8)$$

with a constant, symmetric and positive definite matrix κ^* *given by (3.4).*

Proof. We proceed along the lines of the proof presented in the recent work [137] by Piiroinen and the author: Let us first show that for $\mathcal{P}$-a.e. $\omega \in \Gamma$, $u_\varepsilon(x,\omega) \to u^*(x)$, $x \in \overline{D}$, as $\varepsilon \to 0$. Let $(\varepsilon_k)_{k \in \mathbb{N}}$ be an arbitrary monotone decreasing null sequence and let X^{ω,ε_k} denote the reflecting diffusion process on the half-space corresponding to the regular symmetric Dirichlet form $(\mathcal{E}^{\omega,\varepsilon_k}, H^1(\mathbb{R}^d_- \cup \mathbb{R}^{d-1}))$ on $L^2(\mathbb{R}^d_- \cup \mathbb{R}^{d-1})$. By assumption

(A1) from Subsection 1.3.1, we deduce from [33, Section 4] that for $\mathcal{P}$-a.e. $\omega \in \Gamma$

$$X^{\omega,\varepsilon_k} \to X^* \quad \text{in law on } C([0,\infty); \mathbb{R}^d_- \cup \mathbb{R}^{d-1}), \text{ as } k \to \infty,$$

where X^* is the reflecting diffusion process on the half-space associated with the homogenized regular symmetric Dirichlet form $(\mathcal{E}^*, H^1(\mathbb{R}^d_- \cup \mathbb{R}^{d-1}))$ on $L^2(\mathbb{R}^d_- \cup \mathbb{R}^{d-1})$. The constant, symmetric and positive definite matrix κ^* is given by (3.4).

Let us first show that for every $x \in \overline{D}$ and $\mathcal{P}$-a.e. realization $\omega \in \Gamma$ of the random medium

$$X^{\omega,\varepsilon_k}_{\cdot \wedge \tau^{\omega,\varepsilon_k}} \to X^*_{\cdot \wedge \tau^*} \quad \text{in law on } C([0,\infty); \mathbb{R}^d_- \cup \mathbb{R}^{d-1}), \text{ as } k \to \infty.$$

Consider the functional $F : C([0,\infty); \mathbb{R}^d_- \cup \mathbb{R}^{d-1}) \to [0,\infty]$,

$$\phi \mapsto \begin{cases} \infty, & \text{if for all } t \geq 0 : |\phi(t)| < R, \\ \inf\{t \geq 0 : |\phi(t)| = R\} & \text{else}, \end{cases}$$

defined in such a way that $F(X^{\omega,\varepsilon_k}) = \tau^{\omega,\varepsilon_k}$. Let $(\phi_k)_{k \in \mathbb{N}}$ denote a sequence of continuous functions that converges uniformly towards ϕ on compacts in $[0,\infty)$. If $\liminf_{k \to \infty} F(\phi_k)$ is finite, then we may extract a subsequence, still denoted $(\phi_k)_{k \in \mathbb{N}}$ for convenience, such that $F(\phi_k) \to \liminf_{k \to \infty} F(\phi_k)$, as $k \to \infty$. We have

$$|\phi(\liminf_{k \to \infty} F(\phi_k)) - \phi_k(F(\phi_k))| \leq |\phi(\liminf_{k \to \infty} F(\phi_k)) - \phi(F(\phi_k))|$$
$$+ |\phi(F(\phi_k)) - \phi_k(F(\phi_k))|$$

and by our assumption, the right-hand side vanishes as $k \to \infty$. From the closedness of $\partial_1 D$, we conclude hence that $\phi(\liminf_{k \to \infty} F(\phi_k)) \in \partial_1 D$. In particular we have that $F(\phi) \leq \liminf_{k \to \infty} F(\phi_k)$ if $\sup_{t \in [0,T]} |\phi(t) - \phi_k(t)| \to 0$, as $t \to \infty$, for all $T > 0$, i.e., F is lower semi-continuous. Now assume that ϕ is a discontinuity point of F, i.e., there is $\delta > 0$ and k_0 such that $F(\phi) + \delta \leq F(\phi_k) < \infty$ for all $k \geq k_0$. Then it follows that $\phi(t) \in \overline{D}$ for all $t \in [F(\phi), \liminf_{k \to \infty} F(\phi_k))$. However, the boundary $\partial_1 D$ is regular in the sense of [80, Chapter 4.2], that is, a diffusion process originating from $x \in \partial_1 D$ will immediately exit from $\overline{D}$ $\mathbb{P}^{\omega,\varepsilon_k}_x$-a.s., respectively $\mathbb{P}^*_x$-a.s. In other words, the set of discontinuities of F is a null set with respect to these measures and hence the claim follows from the continuous mapping theorem, cf. [21]. Now, as in the proof of [145, Theorem 5.1], it follows with the Fukushima decompositions in Section 2.2 of the second chapter that

$$(X^{\omega,\varepsilon_k}_{\cdot \wedge \tau^{\omega,\varepsilon_k}}, L^{\omega,\varepsilon_k}_{\cdot \wedge \tau^{\omega,\varepsilon_k}}) \to (X^*_{\cdot \wedge \tau^*}, L^*_{\cdot \wedge \tau^*}) \quad \text{in law on } C([0,\infty); \mathbb{R}^d_- \cup \mathbb{R}^{d-1} \times \mathbb{R}_+),$$

as $k \to \infty$. With regard to the Feynman-Kac formula (2.38), the assertion (3.5) can be proved as follows:

Suppose first that f is continuous. Then we show that

$$\limsup_{j \to \infty} \left| \mathbb{E}_x \int_0^{\tau_j} e_g^{\omega,\varepsilon_j}(t) f(X_t^{\omega,\varepsilon_j}) \, \mathrm{d}L_t^{\omega,\varepsilon_j} - \mathbb{E}_x \int_0^{\tau} e_g(t) f(X_t) \, \mathrm{d}L_t \right| = 0$$

by estimating with a difference of truncated Riemann sums

$$\left| \sum_{k=0}^{\lfloor N/h \rfloor} \left(\mathbb{E}_x \int_{t_k}^{t_{k+1}} e_g^{\omega,\varepsilon_j}(t_k) f(X_{t_k}^{\omega,\varepsilon_j})[t_{k+1} < \tau_j] \, \mathrm{d}L_t^{\omega,\varepsilon_j} \right. \right.$$

$$\left. \left. - \mathbb{E}_x \int_{t_k}^{t_{k+1}} e_g(t_k) f(X_{t_k})[t_{k+1} < \tau] \, \mathrm{d}L_t \right) \right|,$$

where $t_k = kh$ and h is a step size. The difference $S_{N,h}$ of truncated Riemann sums for fixed N and h goes to zero as $j \to \infty$ if we assume that $(X^{\omega,\varepsilon_j}, L^{\omega,\varepsilon_j}, A^{\omega,\varepsilon_j})$ converge weakly to (X, L, A), as $j \to \infty$, where we have set $A^{\omega,\varepsilon_j} := \log e_g^{\omega,\varepsilon_j}$ and $A := \log e_g$.

The error terms for X^{ω,ε_j} and X may be treated analogously, so it is enough to consider just X^{ω,ε_j} in detail. The increments (excluding the edge case where $t_k < \tau_j < t_{k+1}$ which we will omit but which can be treated in the same way) are of following form

$$\int_{t_k}^{t_{k+1}} \left(e_g^{\omega,\varepsilon_j}(t) f(X_t^{\omega,\varepsilon_j}) - e_g^{\omega,\varepsilon_j}(t_k) f(X_{t_k}^{\omega,\varepsilon_j}) \right) \mathrm{d}L_t^{\omega,\varepsilon_j}$$

$$= \int_{t_k}^{t_{k+1}} \left(e_g^{\omega,\varepsilon_j}(t) - e_g^{\omega,\varepsilon_j}(t_k) \right) f(X_t^{\omega,\varepsilon_j}) \, \mathrm{d}L_t^{\omega,\varepsilon_j}$$

$$+ e_g^{\omega,\varepsilon_j}(t_k) \int_{t_k}^{t_{k+1}} \left(f(X_t^{\omega,\varepsilon_j}) - f(X_{t_k}^{\omega,\varepsilon_j}) \right) \mathrm{d}L_t^{\omega,\varepsilon_j}$$

The latter term can be handled with the continuity of the paths of X together with the uniform continuity of f on the compact set $\overline{D}$. This is since

$$\left| e_g^{\omega,\varepsilon_j}(t_k) \int_{t_k}^{t_{k+1}} \left(f(X_t^{\omega,\varepsilon_j}) - f(X_{t_k}^{\omega,\varepsilon_j}) \right) \mathrm{d}L_t^{\omega,\varepsilon_j} \right|$$

$$\leq \left(2 \left(1 - \psi_\delta(\theta_h(X^{\omega,\varepsilon_j})) \right) \|f\|_\infty + \theta_{2\delta}(f) \psi_\delta(\theta_h(X^{\omega,\varepsilon_j})) \right) (L_{t_{k+1}}^{\omega,\varepsilon_j} - L_{t_k}^{\omega,\varepsilon_j}),$$

where $\theta_\delta(x)$ is the maximum variation of the function x on the interval $[0, N]$

$$\theta_\delta(x) := \sup \left\{ |x(t) - x(s)| \, ; \, 0 \leq t, s \leq N, |t - s| < \delta \right\}$$

and $\psi_\delta(t)$ is the continuous approximation of the indicator function $[t < \delta]$ with support in $[-2\delta, 2\delta]$. Therefore, after taking the limit $j \to \infty$, the latter terms give that the corresponding total approximation error can be bounded by

$$4\|f\|_\infty \mathbb{E}_x\{[\theta_h(X) \geq \delta]L_{\tau \wedge N}\} + 2\theta_{2\delta}(f)\mathbb{E}_x\{L_{\tau \wedge N}\}$$

which goes to zero for fixed N if we first let $h \to 0$ and then let $\delta \to 0$. The first term can be estimated by

$$\int_{t_k}^{t_{k+1}} \left(e_g^{\omega,\varepsilon_j}(t) - e_g^{\omega,\varepsilon_j}(t_k)\right) f(X_t^{\omega,\varepsilon_j}) \, \mathrm{d}L_t^{\omega,\varepsilon_j}$$

$$\leq 2\|f\|_\infty\|g\|_\infty \int_{t_k}^{t_{k+1}} \left(L_t^{\omega,\varepsilon_j} - L_{t_k}^{\omega,\varepsilon_j}\right) \, \mathrm{d}L_t^{\omega,\varepsilon_j} = \|f\|_\infty\|g\|_\infty (L_{t_{k+1}}^{\omega,\varepsilon_j} - L_{t_k}^{\omega,\varepsilon_j})^2$$

since $g \geq 0$. Therefore, after taking $j \to \infty$ the first term gives the total approximation error that is bounded by

$$2\|f\|_\infty\|g\|_\infty \left(\mathbb{E}_x\{[\theta_h(L) \geq \delta]L_{\tau \wedge N}\} + 2\delta\mathbb{E}_x\{L_{\tau \wedge N}\}\right)$$

which goes to zero for fixed N if we first let $h \to 0$ and then let $\delta \to 0$.

Now, the truncation can be removed since τ_j and τ are a.s. finite and moreover, $L_{\tau_j}^{\omega,\varepsilon_j}$ converges weakly to L_τ, as $j \to \infty$, and thus, we get a uniform estimate

$$\limsup_{j \to \infty} \left|\mathbb{E}_x \int_0^{\tau_j} e_g^{\omega,\varepsilon_j}(t)f(X_t^{\omega,\varepsilon_j}) \, \mathrm{d}L_t^{\omega,\varepsilon_j} - \int_0^\tau e_g(t)f(X_t) \, \mathrm{d}L_t\right|$$

$$\leq 2\|f\|_\infty \mathbb{E}_x\{L_\tau - L_{\tau \wedge N}\}$$

which gives the claimed convergence for continuous f once we have proved weak convergence of A^{ω,ε_j} as $j \to \infty$.

Therefore, we will next verify the weak convergence of A^{ω,ε_j} jointly with $(X^{\omega,\varepsilon_j}, L^{\omega,\varepsilon_j})$. We can assume that $X^{\omega,\varepsilon_j} \to X$ and $L^{\omega,\varepsilon_j} \to L$ almost surely in $C(0,T)$. Suppose that $g = \sum_{l=1}^N g_l[E_l]$ where $g_l \geq 0$ and continuous, $E_l \cap E_k = \emptyset$ and $\sigma(\partial E_l) = 0$. It is enough to show the convergence for $g := [E_l]$. Since E_l is open in ∂D, it can be approximated from below by an increasing sequence of continuous functions g'_k that converge pointwise to g. It follows from the almost sure convergence of X^{ω,ε_j} and L^{ω,ε_j} that

$$\liminf_{j \to \infty} \int_0^t g(X_t^{\omega,\varepsilon_j}) \, \mathrm{d}L_t^{\omega,\varepsilon_j} \geq \limsup_{k \to \infty} \int_0^t g'_k(X_t) \, \mathrm{d}L_t = \int_0^t g(X_t) \, \mathrm{d}L_t.$$

Since X^{ω,ε_j} and X will not hit ∂E_l in $[0, T]$ almost surely, we can get the other direction by considering $1 - g$. Therefore,

$$\limsup_{j\to\infty} \int_0^t g(X_t^{\omega,\varepsilon_j})\,\mathrm{d}L_t^{\omega,\varepsilon_j} \leq \int_0^t g(X_t)\,\mathrm{d}L_t.$$

Using countability of the rational numbers, we obtain

$$\forall t \in [0,T] \cap \mathbb{Q}: \lim_{j\to\infty} A^{\omega,\varepsilon_j}(t) = A(t)$$

almost surely and by monotonicity this implies the almost sure convergence of A^{ω,ε_j} to A in $C(0,T)$, as $j \to \infty$, which implies the weak convergence of the original versions. The same technique allows us to extend the assertion to the case of discontinuous f.

For the proof of (3.6) note that the boundary condition (1.17) allows us to write

$$J_l(\varepsilon_k, \omega) = \frac{1}{|E_l|} \int_{E_l} (f - g u_{\varepsilon_k}(\cdot, \omega)|_{\partial_1 D})\,\mathrm{d}\sigma(x) \quad \text{for } \mathcal{P}\text{-a.e. } \omega \in \Gamma \quad (3.9)$$

and that (1.17) may be written in the form $(\Lambda_{\kappa_\varepsilon} + gI)u_\varepsilon = f$. We deduce from the well-posedness of the forward problem that $(\Lambda_{\kappa_\varepsilon} + gI)^{-1}$, $L^2(\partial D) \to L^2(\partial D)$, is bounded. Since, due to assumption (A1), the corresponding constant does not depend on ε, the sequence $(u_{\varepsilon_k})_{k\in\mathbb{N}}$ is bounded in $L^2(\partial D)$ for $\mathcal{P}$-a.e. $\omega \in \Gamma$ implying its uniform integrability, see, e.g., [88]. As we already know the pointwise convergence $u_{\varepsilon_k}(x, \omega) \to u^*(x)$, $x \in \overline{D}$, as $k \to \infty$, an application of Egorov's theorem yields convergence in $L^1(\partial D)$ so that the assertion follows by the triangle inequality and taking the limit $k \to \infty$ inside the integration in (3.9). $\qquad\square$

Remark 3.3. We would like to point out that the effective conductivity κ^* is determined by the invariance principle on the whole space $\mathbb{R}^d$. That is, the apparently non-standard problem of homogenizing the stochastic EIT forward problem (1.16), (1.17) which seemingly requires homogenization of a reflecting diffusion process is transformed into a standard one, namely the homogenization of the corresponding diffusion process on $\mathbb{R}^d$.

3.3 A continuum micro-scale Monte Carlo method

Formula (3.4) is of no immediate use for practical computations since it is formulated on the whole space $\mathbb{R}^d$ and has to be solved for $\mathcal{P}$-a.e. realization of the random medium. While the latter issue can be overcome (at least theoretically) thanks to the ergodicity of the underlying dynamical system, the problem of choosing an appropriate spatial cut-off with appropriate boundary conditions is more severe. Still, Monte Carlo methods based on finite element discretization of the auxiliary problem (3.2) are widely considered as a state-of-the-art tool for numerical homogenization.

Theorem 3.2 from the previous section motivates an alternative and rather different approach which is worthwhile exploring, namely the discretization of the underlying diffusion process. This idea is clearly inspired by the work [159], where a stochastic method based on simulation of the Brownian motion is introduced which is indeed considered to be rather accurate, cf. Karim and Krabbenhoft [81].

In this section, we derive a new stochastic numerical homogenization method for the approximation of the effective conductivity κ^* which is based on simulation of paths of the underlying diffusion process. Moreover, we provide a rigorous convergence analysis for this method.

3.3.1 A note on discrete random walk methods

The discrete lattice random walk in random environments is well studied, both in the physics and the mathematics literature and there are numerous recent publications devoted to stochastic homogenization methods for discrete difference equations, cf., e.g., Egloffe, Gloria, Mourrat and Nguyen [43] or Gloria and Mourrat [56]. Therefore, one might be tempted to use such a method in the continuum setting as well. However, it is known that the peculiarities of (digitized) continuum media caused, e.g., by narrow "necks" or corners that join two conducting phase pixels, are not captured by lattice random walks, cf., e.g., [83]. Consider, for instance, the standard two-dimensional periodic checkerboard with cells of unit length and conductivities κ_1 and κ_2, respectively. It is well known that in this case the isotropic effective conductivity is given by $\kappa^* = \sqrt{\kappa_1 \kappa_2}$. In [83], a discrete random walk on lattice points given by the centers of the cells is considered and it is shown that the approximation for the effective conductivity from this random walk is given by the harmonic mean $\hat{\kappa}^* = 2\kappa_1\kappa_2(\kappa_1 + \kappa_2)^{-1}$. This estimate is way off if the conductivity contrast κ_1/κ_2 differs significantly from unity. Of course, the random walk may be seen as a randomized version of the finite difference method so that one could be tempted to increase the spatial resolution of the lattice. The question how far the effective coefficients of such a difference scheme differ from those of the corresponding continuum differential operators was first studied by Avellaneda, Hou and

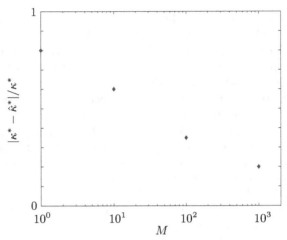

Figure 3.1: Random walk effective conductivity approximation for the periodic checkerboard with $\kappa_1 = 1$, $\kappa_2 = 100$. Each cell is discretized by an $M \times M$ mesh. The relative error is plotted against M.

Papanicolaou [9], where it was shown that, in the multidimensional case, the finite difference approach does not provide the correct effective coefficients unless the ratio of the size of the discretization mesh to the microscopic length scale goes to zero. For an illustration of this theoretical result see Figure 3.1, where the relative errors of a random walk approximation of the effective conductivity for the periodic checkerboard are plotted for different sizes of the discretization mesh. To keep the random error reasonably small, we used 10^6 random walks. Finally, we would like to point out that it has been shown by Piatnitski and Remy that the effective coefficients for discrete difference equations depend essentially on the discretization method [135].

3.3.2 Path simulation in digitized random media

There exists now a vast variety of microscopic models that have been proposed to represent the micro-scale structure of heterogeneous media, cf., e.g., the monograph by Torquato [158]. In practice, samples of such media are commonly represented as *digitized media*; these are obtained by tiling the space with tiles of size of the order of magnitude of the grain size of the heterogeneous medium. Examples of such digitized media include honeycomb-like tilings or checkerboard-like tilings, see Figure 3.2. For a rigorous construction of the corresponding probability space $(\Gamma, \mathcal{G}, \mathcal{P})$ and

Figure 3.2: Checkerboard-like digitized random media.

the underlying dynamical system we refer the reader to the recent work by
Alexanderian, Rathinam and Rostamian [4].

For simplicity, we restrict ourselves here to the case of checkerboard-
like media and without loss of generality, we assume that $d = 2$. For
the description of the path simulation scheme, we consider a rectangular
model domain $D = [-2a, 2a] \times [-a, a]$ which is divided into two tiles $D_1 =
[-2a, 0] \times [-a, a]$ and $D_2 = (0, 2a] \times [-a, a]$ such that

$$\kappa(x) := \begin{cases} \kappa_1, & x \in D_1 \\ \kappa_2, & x \in D_2 \end{cases}$$

with constants $\kappa_1, \kappa_2 > 0$. We have shown in Chapter 2, that in D the
corresponding diffusion process starting in $(x, y)^T$ is then given by the
solution to the following system of one-dimensional stochastic differential
equations

$$\begin{cases} X_t &= x + \int_0^t \sqrt{2\kappa((X_s, Y_s)^T)}\, dW_s^{(1)} + \frac{\kappa_1 - \kappa_2}{\kappa_1 + \kappa_2} L_t^0(X) \\ Y_t &= y + \int_0^t \sqrt{2\kappa((X_s, Y_s)^T)}\, dW_s^{(2)}, \end{cases} \qquad (3.10)$$

where $(W^{(1)}, W^{(2)})^T$ is a standard two-dimensional Brownian motion and
$L^0(X)$ is the symmetric local time of X at 0. In particular, simulation of this
process inside D_1 and D_2 comes down to simulation of the standard Brownian
motion with appropriate scaling. This can be done exactly using the so-called
random walk on squares which we describe in the next paragraph. Moreover,

it can be seen from (3.10) that the interface behavior of the process is governed by the symmetric local time of X at 0. We will study numerical schemes for the simulation of the interface behavior in the next but one paragraph.

Random walk on squares   The random walk on squares was introduced by Milstein and Tretyakov [123] as a generalization of the well-known random walk on spheres [125]. While the latter introduces a bias due to the fact that the particle must be stopped within a small layer in the vicinity of the boundary, cf. [121], both exit time and position from a polygonal domain may be computed exactly via the random walk on squares. The basic idea is to generate iteratively the first exit time and position from a square, as big as possible in the polygonal domain D, which is centered at the previous position of the Brownian motion. As soon as the exit position lies on ∂D, the algorithm stops. This method relies on the use of analytical expressions for the distribution functions of the first exit time and position, which may be deduced from analytical expressions for the one-dimensional fundamental solution to a parabolic initial boundary value problem for the Laplacian.

For convenience of the reader, let us discuss the main ingredients of the method in more detail: The goal is to simulate the first exit time $\tau(S)$ at which the scaled Brownian motion $\sqrt{2\kappa}W$ starting from the center of a square S exits from the latter, together with the position $\sqrt{2\kappa}W_{\tau(S)}$. Without loss of generality, we may assume that $\kappa = 1/2$ and that $S = (-1, 1)^2$.

Let us first study the one-dimensional situation. We denote by τ the first exit time from the interval $(-1, 1)$ of the one-dimensional Brownian motion starting in x. By the standard Feynman-Kac formula, we know that the function $v(t, x) := \mathbb{P}_x\{\tau < t\}$ is the unique solution to the parabolic initial boundary value problem

$$\partial_t v = \frac{1}{2}\partial_{xx}v, \quad t > 0, \quad x \in (-1, 1) \tag{3.11}$$

subject to the initial and boundary conditions

$$v(0, x) = 0, \quad x \in [-1, 1]$$
$$v(t, x) = 1, \quad t > 0, \ x \in \{-1, 1\}.$$

In particular, the function $w(t, x) := v(t, x) - 1$ satisfies (3.11) subject to the initial and boundary conditions

$$w(0, x) = -1, \quad x \in [-1, 1]$$
$$w(t, x) = 0, \quad t > 0, \ x \in \{-1, 1\}.$$

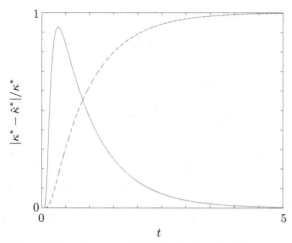

Figure 3.3: Distribution function $F(t)$ (dashed line) of the first exit time from the interval $(-1, 1)$ of the standard Brownian motion and the corresponding density $\dot{F}(t)$ (solid line).

Using separation of variables for this problem, we obtain the expression

$$\mathbb{P}_x\{\tau < t\} = 1 - \frac{4}{\pi} \sum_{k=0}^{\infty} \frac{(-1)^k}{2k+1} \cos\left(\frac{1}{2}\pi(2k+1)x\right) \exp\left(-\frac{1}{8}\pi^2(2k+1)^2 t\right).$$

On the other hand, the transition density kernel of the Brownian motion on $(-1, 1)$ may be computed explicitly using the method of images. Namely, $p(t, x, y)$ is given by the expression

$$\frac{1}{\sqrt{2\pi t}} \sum_{k=-\infty}^{\infty} \left(\exp\left(-\frac{1}{2t}(x - y - 4k)^2\right) - \exp\left(-\frac{1}{2t}(x + y - 2 - 4k)^2\right) \right),$$

and we may write

$$\mathbb{P}_x\{\tau < t\} = 1 - \int_{-1}^{1} p(t, x, y)\, \mathrm{d}y.$$

Setting $x = 0$, we obtain thus for the distribution function F of τ the equivalent formulae

$$F(t) = 1 - \frac{4}{\pi} \sum_{k=0}^{\infty} \frac{(-1)^k}{2k+1} \exp\left(-\frac{1}{8}\pi^2(2k+1)^2 t\right), \quad t > 0, \qquad (3.12)$$

respectively

$$F(t) = 2 \sum_{k=0}^{\infty} (-1)^k \operatorname{erfc}\left(\frac{2k+1}{\sqrt{2t}}\right), \quad t > 0, \tag{3.13}$$

where erfc is the complementary errorfunction, i.e.,

$$\operatorname{erfc}(x) = \frac{2}{\sqrt{\pi}} \int_x^{\infty} \exp(-y^2) \, \mathrm{d}y.$$

With regard to numerical computations, it can be said that (3.12) is more suitable for large values of t, while (3.13) should be used if t is small, see Figure 3.3 for a plot of both F and the corresponding density $\dot{F}$. Given these expressions for the distribution function, we may simulate realizations of τ via the *inversion method*, i.e., if $\gamma \sim \mathcal{U}[0,1]$, then τ has the distribution $F^{-1}(\gamma)$.

Now, still considering the one-dimensional standard Brownian motion W, we have by Bayes' formula

$$\mathbb{P}_0\{W_t < z \,||W_s| < 1 \, \forall s \in (0,t)\} = \mathbb{P}_0\{W_t < z | \tau \geq t\}$$
$$= \mathbb{P}_0\{W_t < z, \tau \geq t\} \cdot (\mathbb{P}_0\{\tau \geq t\})^{-1}$$

and we denote this conditional probability $\mathcal{Q}_t(z)$. From the right-hand side of the last equality together with the expressions (3.13) and (3.12) one obtains by straightforward computations, cf. [123, Lemma 3.2], that

$$\mathcal{Q}_t(z) = (1 - F(t))^{-1} \frac{2}{\pi} \sum_{k=0}^{\infty} \frac{1}{2k+1} \left((-1)^k + \sin\left(\frac{1}{2}\pi(2k+1)z\right) \right)$$
$$\times \exp\left(-\frac{1}{8}\pi^2 (2k+1)^2 t \right), \tag{3.14}$$

respectively,

$$\mathcal{Q}_t(z) = (1 - F(t))^{-1} \sum_{k=0}^{\infty} \frac{(-1)^k}{2} \left(\operatorname{erfc}\left(\frac{2k-1}{\sqrt{2t}}\right) - \operatorname{erfc}\left(\frac{2k+z}{\sqrt{2t}}\right) \right.$$
$$\left. - \operatorname{erfc}\left(\frac{2k+2-z}{\sqrt{2t}}\right) + \operatorname{erfc}\left(\frac{2k+3}{\sqrt{2t}}\right) \right). \tag{3.15}$$

Again, with regard to numerical computations, the expression (3.14) is more suitable for large values of t, whereas (3.15) should be preferred if t is small.

Algorithm 1 Exact simulation of exit time $\tau(S)$ and position $\sqrt{2\kappa}W_{\tau(S)}$ of the scaled Brownian morion from square S centered at $(0,0)^T$

Require: edge length $2a$, κ
Draw $\gamma_1 \sim \mathcal{U}[0,1]$, $\gamma_2 \sim \mathcal{U}[0,1]$
Calculate $\tau \leftarrow F^{-1}(1 - \sqrt{\gamma_1})$, $z \leftarrow \mathcal{Q}_\tau^{-1}(\gamma_2)$
Draw $s \sim \mathcal{U}\{\pm 1\}$, $i \sim \mathcal{U}\{1,2\}$
Set $W^{(i)} \leftarrow s$, $W^{(2-(i-1))} \leftarrow z$
Set $\tau(S) \leftarrow a^2\sqrt{2}\kappa^{-1}\tau$, $\sqrt{2\kappa}W_{\tau(S)} \leftarrow aW$
return $\tau(S)$, $\sqrt{2\kappa}W_{\tau(S)}$

Let us now combine these results for the one-dimensional Brownian motion to obtain an algorithm for the simulation of $(\tau(S), W_{\tau(S)})$, where S is a square with edge length 2 centered at $(0,0)^T$. First, note that the two-dimensional distribution function $\mathbb{P}_0(\tau(S) < t)$ is given by $1 - (1 - F(t))^2$ so that if $\gamma \sim \mathcal{U}[0,1]$, then $\tau(S)$ has the distribution $F^{-1}(1-\sqrt{\gamma})$. Moreover, we exploit the fact that the components of the Brownian motion are independent, that is, if the boundary of S is first hit, say by $W^{(1)}$, at time τ, then we know that the component $W^{(2)}$ must have been inside $(-1,1)$ during the time interval $[0,\tau]$. That is, to simulate $W_{\tau(S)}$, we can choose one of the edges of the square with equal probability and then compute the exit position on this particular edge using a one-dimensional standard Brownian motion. The exact simulation of $(\tau(S), \sqrt{2\kappa}W_{\tau(S)})$ for a square S with edge length $2a$ centered at $(0,0)^T$ is summarized in Algorithm 1.

Remark 3.4. For the sake of efficiency, we precompute the exit times and corresponding positions from a square so that in actual simulations we can draw them randomly from a list. Note also that, depending on the random medium at hand, the *random walk on rectangles* which has been proposed by Deaconu and Lejay [38] should be considered a valuable alternative to the random walk on squares.

Simulation of the interface behavior

In the last decade, the numerical simulation of diffusion processes with discontinuous coefficients has been the subject of active research, cf., e.g., [25, 45, 46, 47, 106, 108, 110, 116, 117, 120, 122]. While the one-dimensional situation is well understood, the question whether there exist exact simulation schemes for the multi-dimensional case is still unsettled. In this paragraph, we consider the simulation in the special case of digitized media. More precisely, we consider the *mixing scheme* by Lejay and Maire [108] which is based on exact simulation of excursions of the one-dimensional *skew Brownian motion.*

The mixing scheme by Lejay and Maire Consider the one-dimensional stochastic differential equation

$$X_t = x + \int_0^t \sqrt{2\kappa(X_s)}\, dW_s + \frac{\kappa_1 - \kappa_2}{\kappa_1 + \kappa_2} L_t^0(X) \qquad (3.16)$$

with discontinuous coefficient

$$\kappa(x) := \begin{cases} \kappa_1, & x \le 0 \\ \kappa_2, & x > 0, \end{cases}$$

$\kappa_1, \kappa_2 > 0$, and $L^0(X)$ the symmetric local time of X at 0. In order to reduce (3.16) to the so-called *skew Brownian motion*, we use the one-to-one mapping $\psi(x) := x/\sqrt{2\kappa(x)}$. It follows from the Tanaka formula and the properties of the symmetric local time that $Z_t := \psi(X_t)$ is the solution to the stochastic differential equation

$$Z_t = \psi(x) + W_t + \theta L_t^0(Z), \quad \theta := \frac{\sqrt{\kappa_1} - \sqrt{\kappa_2}}{\sqrt{\kappa_1} + \sqrt{\kappa_2}}. \qquad (3.17)$$

In fact, the solution Z to (3.17) is the *skew Brownian motion* with parameter $\alpha = \frac{\theta+1}{2} \in (0,1)$. This process, which was introduced by Itô and McKean [74], behaves like the reflecting Brownian motion with reflecting boundary at 0, except that the sign of each excursion is chosen using an independent Bernoulli random variable of parameter α. That is, with probability α the excursion has a positive sign, otherwise a negative one. Due to this construction of the skew Brownian motion, we can simulate excursions of the one-dimensional process (3.16) exactly, see Algorithm 2 below. See also the survey article [105] by Lejay for a concise review of various equivalent constructions of the skew Brownian motion. Another possible construction uses transition density kernels. The transition kernel density p of X and the transition kernel density p_θ of Z are linked via the relation

$$p(t,x,y) = \frac{1}{\sqrt{\kappa}} p_\theta\left(t, \frac{x}{\sqrt{\kappa}}, \frac{y}{\sqrt{\kappa}}\right),$$

where

$$p_\theta(t,x,y) = p_{\mathcal{G}}(t, y - x) + \text{sign}(y)\theta p_{\mathcal{G}}(t, |x| + |y|).$$

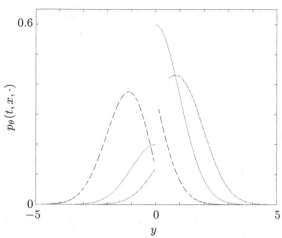

Figure 3.4: Transition kernel density of the skew Brownian motion for $\theta = 0.5$ at time $t = 1$ evaluated for the initial points $x = -1$ (dashed line), 0 (solid line), 1 (dash-dotted line).

Here, we have assumed that $\text{sign}(0) = 1$, and $p_{\mathcal{G}}$ denotes the Gaussian density

$$p_{\mathcal{G}}(t, x) = \frac{1}{2\pi t} \exp\left(\frac{-x^2}{2t}\right).$$

Plots of the transition kernel density p_θ for different values of the initial point x can be found in Figure 3.4. In order to define a proper interface

---

Algorithm 2 Simulation of $\widehat{X}_t$, given $X_0 = 0$, via skew Brownian motion

---

Require: timestep h, κ
  Set $t \leftarrow h$
  Calculate $\alpha \leftarrow \sqrt{\kappa_2}/(\sqrt{\kappa_1} + \sqrt{\kappa_2})$
  Draw $\gamma_1 \sim \mathcal{U}[0, 1]$, $\gamma_2 \sim \mathcal{N}(0, 1)$
  if $\gamma_1 \leq \alpha$ **then**
    Set $\widehat{X}_t \leftarrow \sqrt{2\kappa_2 h}|\gamma_2|$
  else
    Set $\widehat{X}_t \leftarrow -\sqrt{2\kappa_1 h}|\gamma_2|$
  end if
  return X_t, t

---

scheme, it remains to simulate the behavior of the second component which
is the solution to the stochastic differential equation

$$Y_t = y + \int_0^t \sqrt{2\kappa((X_s, Y_s)^T)} \, dW_s^{(2)}. \tag{3.18}$$

The following approximation of (3.18) to simulate $Y_{t+\delta t}$ from Y_t was proposed
in [108]: Since the processes X and $W^{(2)}$ are independent, the law of Y_t
conditioned on $Y_0 = y$ and X follows a Gaussian distribution with mean y
and variance

$$2 \int_0^t \kappa(X_s) \, ds = 2(\kappa_1 A_1(t) + \kappa_2 A_2(t)),$$

where

$$A_1(t) = \int_0^t [X_s \leq 0] \, ds \quad \text{and} \quad A_2(t) = \int_0^t [X_s \geq 0] \, ds$$

denote the *occupation times* of $\mathbb{R}^+$ and $\mathbb{R}^-$ for X, respectively. For $x = 0$,
we know by the properties of the skew Brownian motion that

$$\mathbb{E}_0 A_1(t) = t(1 - \alpha) \quad \text{and} \quad \mathbb{E}_0 A_2(t) = t\alpha.$$

The y-component is now approximated by a Gaussian random variable with
variance

$$2\mathbb{E}_0 \int_0^t \kappa(X_s) \, ds,$$

that is,

$$\widehat{Y}_t = y + \sqrt{2t(\kappa_1(1 - \alpha) + \kappa_2\alpha)}\gamma, \quad \gamma \sim \mathcal{N}(0, 1). \tag{3.19}$$

Algorithm 3 describes the mixing scheme for the simulation of the interface
behavior.

Algorithm 3 Simulation of $(\widehat{X}_t, \widehat{Y}_t)^T$, given $(X_0, Y_0)^T = (0, y)^T$, via mixing
scheme

Require: timestep h, κ
  Set $t \leftarrow h$
  Calculate $\alpha \leftarrow \sqrt{\kappa_2}/(\sqrt{\kappa_1} + \sqrt{\kappa_2})$
  Simulate $\widehat{X}_t$ according to Algorithm 2
  Draw $\gamma \sim \mathcal{N}(0, 1)$
  Set $\widehat{Y}_t \leftarrow (0, y)^T + \sqrt{2h(\kappa_1(1 - \alpha) + \kappa_2\alpha)}\gamma$
  return $(\widehat{X}_t, \widehat{Y}_t)^T$, t

Remark 3.5. Note that it would be possible to simulate the occupation times exactly, see, e.g., [24] for the analytical expressions of the corresponding probability densities. However, in order to derive an exact simulation scheme for the case of digitized media, it remains an open question how to simulate exactly these occupation times conditioned on $Y_0 = y$ and X.

Remark 3.6. An alternative approach for the simulation of the interface behavior could use the fact that excursions of the multi-dimensional skew Brownian motion may be simulated exactly. In fact, one could simply implement the definition of the skew Brownian motion due to Itô and McKean [74] as follows: Simulate a process that behaves like the reflecting Brownian motion with reflection at the interface except that the sign of each excursion is chosen using an independent Bernoulli random variable of parameter α. These excursions coincide locally with the reflecting Brownian motion in the half-space which can be simulated exactly as has been pointed out by Lépingle [111]. To be precise, the boundary local time of the (scaled) reflecting Brownian motion $X_t = x + \sqrt{2\kappa}W_t + \nu L_t$ on the half-space can be computed explicitly from the Skorohod problem

$$L_t = \max\left\{0, \sup_{0 \le s \le t}\left\{-\sqrt{2\kappa}\nu \cdot W_s\right\}\right\}, \quad t \ge 0.$$

In particular, the pair (X_t, L_t) may be simulated exactly: Let $\gamma_1 \sim \mathcal{N}(0, tI_2)$ and $\gamma_2 \sim \text{Exp}(1/2t)$, then a realization of $\left(W_t, \sup_{0 \le s \le t}\{-\sqrt{2\kappa}\nu \cdot W_s\}\right)$ is given by the pair

$$\left(\gamma_1, \frac{1}{2}\left(-\sqrt{2\kappa}\nu \cdot \gamma_1 + \sqrt{|-\sqrt{2\kappa}\nu|^2\gamma_2 + (-\sqrt{2\kappa}\nu \cdot \gamma_1)^2}\right)\right).$$

In our numerical studies, the empirical densities of the joint exit distributions from a square for both the time and the position obtained by this method strongly resembled those obtained by the mixing scheme. However, for the sake of brevity, we postpone the further development as well as the theoretical analysis of this alternative approach to future research.

Homogenization method

Henceforth, let X denote the two-dimensional diffusion process on $\mathbb{R}^2$ and let $\mathcal{I}$ denote the *skeleton* of tile interfaces for the checkerboard tiling. A combined simulation scheme for the discretization of X based on Algorithms 1 and 3 is presented in Algorithm 4. We would like to point out that in practice, we proceed as follows: We precompute an empirical density of the

---

Algorithm 4 Simulation of $\widehat{X}_T$ for a fixed time horizon T.

Require: Time horizon T, κ
　Set $t \leftarrow 0$, $\widehat{X}_t \leftarrow (0,0)^T$
　while $t < T$ **do**
　　if $\widehat{X}_t \in \mathcal{I}$ **then**
　　　Simulate $\widetilde{X}_h$ according to Algorithm 3 with (adaptive) timestep h
　　　chosen small enough to ensure that only one interface is crossed;
　　　Set $\widehat{X}_{t+h} \leftarrow \widehat{X}_t + \widetilde{X}_h$, $t \leftarrow t + h$
　　else
　　　Set $S \leftarrow$ largest possible square inside current tile centered at $\widehat{X}_t$;
　　　Simulate $(\tau(S), \sqrt{2\kappa(\widehat{X}_t)}W_{\tau(S)})$ according to Algorithm 1;
　　　Set $\widehat{X}_{t+\tau(S)} \leftarrow \widehat{X}_t + \sqrt{2\kappa(\widehat{X}_t)}W_{\tau(S)}$, $t \leftarrow t + \tau(S)$;
　　end if
　end while
　return $\widehat{X}_t$

---

joint exit distribution from a square divided by the interface. Then, as for the walk on squares, we draw samples from this density.

We propose to approximate the projected effective conductivity via double randomization over realizations of both the random medium and the stochastic process evolving in this medium. More precisely, we want to approximate the projected mean-square displacement averaged with respect to the annealed measure $\overline{\mathbb{P}}$, that is,

$$\overline{\mathbb{E}}(X_T \cdot \xi)^2 = \mathbb{M}\mathbb{E}_0(X_T^\omega \cdot \xi)^2, \qquad (3.20)$$

where $\xi \in \mathbb{R}^2$ is a fixed direction vector. Note that X is now defined on the product probability space $\overline{\Omega}$. Clearly, (3.20) remains true if we substitute X_T on the left-hand side with an unbiased estimator $\widehat{X}_T$, implying that in principle it is sufficient to construct only one realization of $\widehat{X}_T^\omega$ for each realization ω of the random medium.

With regard to the variance of the resulting numerical estimator, it may be worthwhile to use the slightly more general estimator

$$\overline{\mathbb{E}}(X_T \cdot \xi)^2 \approx M_1^{-1}M_2^{-1}\sum_{j=1}^{M_1}\sum_{k=1}^{M_2}(\widehat{X}_T^{\omega(j)}(k) \cdot \xi)^2$$

with positive numbers of samples M_1 and M_2. To obtain an estimator for the projected effective conductivity, we have to rescale which yields the biased estimator

$$\xi \cdot \hat{\kappa}_T^*(M_1, M_2)\xi := \frac{1}{2T} M_1^{-1} M_2^{-1} \sum_{j=1}^{M_1} \sum_{k=1}^{M_2} (\widehat{X}_T^{\omega^{(j)}}(k) \cdot \xi)^2. \tag{3.21}$$

3.3.3 Convergence analysis

In order to carry out the convergence analysis, we denote the expectation and variance with respect to the corresponding approximating measure generated by Algorithm 4 by $\widehat{\mathbb{E}}^{M_1,M_2}$, respectively $\widehat{\mathbb{V}}^{M_1,M_2}$, and for the sake of readability, we suppress the dependance of $\hat{\kappa}_T^*(M_1, M_2)$ on M_1 and M_2 for arguments of $\widehat{\mathbb{E}}^{M_1,M_2}$ and $\widehat{\mathbb{V}}^{M_1,M_2}$. The mean square error

$$\widehat{\mathbb{E}}^{M_1,M_2}(\xi \cdot \hat{\kappa}_T^*\xi - \xi \cdot \kappa^*\xi)^2$$

of the estimator (3.21) can be decomposed in the form

$$\widehat{\mathbb{E}}^{M_1,M_2} \left(\xi \cdot \hat{\kappa}_T^*\xi - \widehat{\mathbb{E}}^{M_1,M_2}\xi \cdot \hat{\kappa}_T^*\xi + \widehat{\mathbb{E}}^{M_1,M_2}\xi \cdot \hat{\kappa}_T^*\xi - \xi \cdot \kappa^*\xi \right)^2.$$

This can be equivalently written as

$$\widehat{\mathbb{V}}^{M_1,M_2}\xi \cdot \hat{\kappa}_T^*\xi + \left(\widehat{\mathbb{E}}^{M_1,M_2}\xi \cdot \hat{\kappa}_T^*\xi - \frac{\overline{\mathbb{E}}(X_T \cdot \xi)^2}{2T} + \frac{\overline{\mathbb{E}}(X_T \cdot \xi)^2}{2T} - \xi \cdot \kappa^*\xi \right)^2, \tag{3.22}$$

where the first term, the so-called *Monte Carlo error*, is due to the variance of the estimator and the second term is the square of the sum of the *bias* due to the weak discretization error of the simulation scheme and the bias due to running the simulation only for a finite time horizon T.

For the sake of simplicity, we neglect in our analysis the weak discretization error, i.e., we assume

$$\widehat{\mathbb{E}}^{M_1,M_2}\xi \cdot \hat{\kappa}_T^*\xi - \frac{\overline{\mathbb{E}}(X_T \cdot \xi)^2}{2T} \approx 0.$$

In fact, in our numerical experiments, for reasonable values of T, we could not even observe the weak discretization error since it was dominated by the bias due to the finite time horizon.

However, even under this simplifying assumption, a rigorous convergence analysis for this method requires a quantitative estimate that is stronger than the qualitative result (3.3), which was obtained in [104] using merely the central limit theorem for martingales. We proceed along the lines of recent work [137] by Piiroinen and the author, namely in the following theorem we provide such a quantitative result by generalizing a classical argument due to Kipnis and Varadhan [86]. The proof relies on new spectral bounds which were obtained recently by Gloria and Otto [58]. We refer the reader to the recent papers [43, 56] for an analogous estimate for the discrete lattice random walk in random environment as well as to the paper [124] which was the first one to use the Kipnis and Varadhan argument in order to obtain quantitative results.

Theorem 3.7. *Let* $\{\kappa_\varepsilon(x,\omega), (x,\omega) \in \mathbb{R}^d \times \Gamma\}$ *be a stationary random field satisfying assumptions (A1) and (A2) from Subsection 1.3.1. Then for every direction vector* $\xi \in \mathbb{R}^d$ *there exist positive constants* c_1, c_2 *such that*

$$\left| \frac{\overline{\mathbb{E}}(X_t \cdot \xi)^2}{2t} - \xi \cdot \kappa^* \xi \right| \leq c_1 \begin{cases} |\log t|^{c_2} t^{-1} & d = 2 \\ t^{-1} & d = 3. \end{cases} \tag{3.23}$$

Proof. For fixed $\omega \in \Gamma$ let us consider the diffusion process X^ω on $\mathbb{R}^d$ which is associated with the symmetric regular Dirichlet form $(\mathcal{E}^{\omega,1}, H^1(\mathbb{R}^d))$ on $L^2(\mathbb{R}^d)$ under the measure $\mathbb{P}_0^\omega$. Following [86], we search for a decomposition of the form

$$X_t^\omega = M_t^\omega + R_t^\omega, \tag{3.24}$$

where M_t^ω is a $\mathbb{P}_0^\omega$-martingale and for every direction $\xi \in \mathbb{R}^d$ the projected remainder $R_t^\omega \cdot \xi$ converges to zero in $L^2(\overline{\Gamma})$ as $t \to \infty$.

Once we have found a suitable decomposition (3.24), we first show that

$$t^{-1}\overline{\mathbb{E}}\{(X_t \cdot \xi)^2 - (M_t \cdot \xi)^2\} = t^{-1}\overline{\mathbb{E}}(R_t \cdot \xi)^2. \tag{3.25}$$

Then we will use spectral calculus to estimate the right-hand side of (3.25).

In order to obtain the decomposition (3.24), we recall that the auxiliary problem (3.2) is equivalent to the following stochastic elliptic equation in the physical space $\mathbb{R}^d$: Find $\chi^\xi \in \mathcal{V}_{\text{pot}}^2(\Gamma)$ such that for $\mathcal{P}$-a.e. $\omega \in \Gamma$, the corresponding potential $\eta^\xi : \mathbb{R}^d \times \Gamma \to \mathbb{R}$ is in $C(\mathbb{R}^d) \cap H^1_{\text{loc}}(\mathbb{R}^d)$ as a function of x, satisfies $\eta^\xi(0,\omega) = 0$ and

$$-\nabla \cdot \kappa(\cdot,\omega)(\xi + \nabla\eta^\xi(\cdot,\omega)) = 0 \quad \text{in } \mathbb{R}^d. \tag{3.26}$$

Let us define the function

$$\phi : \mathbb{R}^d \times \Gamma \to \mathbb{R}^d, \quad \phi(x,\omega) := x + \eta(x,\omega) - \eta(0,\omega),$$

where $\eta := (\eta^{e_1}, ..., \eta^{e_d})^T$ and η^{e_i}, $i = 1, ..., d$, denotes the potential corresponding to the solution to the auxiliary problems for the coordinate directions. As the transition density kernel of X^ω is jointly Hölder-continuous and $\phi_i(\cdot, \omega) \in C(\mathbb{R}^d) \cap H^1_{\mathrm{loc}}(\mathbb{R}^d)$, $i = 1, ..., d$, for $\mathcal{P}$-a.e. $\omega \in \Gamma$, the Fukushima decomposition of $\phi_i(X^\omega_t, \omega)$ holds for every starting point $x \in \mathbb{R}^d$ rather than quasi-every $x \in \mathbb{R}^d$. A straightforward computation using the fact that η is defined via the auxiliary problem yields that the continuous additive functional of zero energy in the Fukushima decomposition vanishes so that

$$\phi_i(X^\omega_t, \omega) = M^{\phi_i(\cdot, \omega)}_t, \quad i = 1, ..., d.$$

We set

$$M^\omega_t := (M^{\phi_1(\cdot, \omega)}_t, ..., M^{\phi_d(\cdot, \omega)}_t)^T \quad \text{and} \quad R^\omega_t := -\eta(X^\omega_t, \omega) + \eta(0, \omega)$$

and consider the quantity

$$\overline{\mathbb{E}}(X_t \cdot \xi)^2 = \overline{\mathbb{E}}(M_t \cdot \xi)^2 + \overline{\mathbb{E}}(R_t \cdot \xi)^2 + 2\overline{\mathbb{E}}(M_t \cdot \xi)(R_t \cdot \xi). \tag{3.27}$$

By computing the predictable quadratic variation of the martingale additive functional we obtain for all $t \geq 0$ and a.e. $x \in \mathbb{R}^d$

$$\mathbb{E}^\omega_x(M^\omega_t \cdot \xi)^2 = \mathbb{E}^\omega_x\left(\int_0^t 2(\xi + \nabla\eta^\xi(X_s, \omega)) \cdot \kappa(X_s, \omega)(\xi + \nabla\eta^\xi(X_s, \omega)) \, ds \right).$$

By the stationarity of $\{\nabla\eta^\xi(x, \omega), (x, \omega) \in \mathbb{R}^d \times \Gamma\}$ with respect to $\mathcal{P}$ we have thus

$$\overline{\mathbb{E}}(M^\omega_t \cdot \xi)^2 = 2\xi \cdot \kappa^* \xi t \quad \text{for all } t \geq 0.$$

Moreover, as in [39], it follows that the last term on the right-hand side of (3.27) vanishes. This can be seen by studying the so-called *environment seen by the particle* process, i.e., the stochastic process defined by

$$Y^\omega_t := \begin{cases} \Theta_{X^\omega_t} \omega & t > 0 \\ \omega & t = 0. \end{cases}$$

$\{Y^\omega_t, t \geq 0\}$ is a stationary process with respect to the annealed measure $\overline{\mathbb{P}}$, i.e., for every finite collection of times $t^{(i)}$, $i = 1, ..., k$, the joint distribution of $Y_{t^{(1)}+h}, ..., Y_{t^{(k)}+h}$ under $\overline{\mathbb{P}}$ does not depend on $h \geq 0$. It is well known that the underlying dynamical system $\{\Theta_x, x \in \mathbb{R}^d\}$ defines a d-parameter group $\{S_x, x \in \mathbb{R}^d\}$ of unitary operators on $L^2(\Gamma)$ by $S_x\psi(\omega) := \psi(\Theta_x\omega)$

and this group is strongly continuous, cf. [104]. Its d infinitesimal generators $(\boldsymbol{D}_1, \mathcal{D}(\boldsymbol{D}_1)), ..., (\boldsymbol{D}_d, \mathcal{D}(\boldsymbol{D}_d))$ are given by

$$\boldsymbol{D}_i \psi = \lim_{h \to 0+} \frac{\boldsymbol{S}_{he_i} \psi - \psi}{h}, \quad i = 1, ..., d,$$

for all $\psi \in L^2(\Gamma)$ such that the limit exists. These operators are closed and densely defined. We denote $\boldsymbol{D} := (\boldsymbol{D}_1, ..., \boldsymbol{D}_d)^T$ and introduce the infinitesimal generator $(\mathcal{L}, \mathcal{D}(\mathcal{L}))$ on $L^2(\Gamma)$ of the environment viewed by the particle process, that is, the non-negative definite self-adjoint operator $\mathcal{L} := -\boldsymbol{D} \cdot \kappa \boldsymbol{D}$ on $L^2(\Gamma)$. Note that due to the fact that the trajectories of the random conductivity field are in general not differentiable, one has to use Dirichlet form theory in order to give a precise meaning to computations involving $\mathcal{L}$. We refer the reader to the work [104] where this has been carried out in some detail. By the self-adjoinedness of $\mathcal{L}$, the law of the environment as viewed from the particle process under $\overline{\mathbb{P}}$ is invariant with respect to time reversal and M^ω is odd by [50, Corollary 2.1], i.e., it changes its sign under time reversal, whereas R^ω, which is the zero energy part of the Fukushima decomposition (3.24), is even by [50, Theorem 2.1]. Thus, we notice that the identity (3.25) holds, as claimed.

We will next show that the estimate (3.23) follows from the following spectral representation

$$\overline{\mathbb{E}}(R_t^i \cdot \xi)^2 = 2 \int_0^\infty (1 - e^{-\lambda t})\lambda^{-2} \, d(E_\lambda \boldsymbol{v}^\xi, \boldsymbol{v}^\xi), \tag{3.28}$$

where $\{E_\lambda, \lambda \in \mathbb{R}\}$ is the unique spectral family given by the spectral theorem such that $\mathcal{L} = \int_0^\infty \lambda \, dE_\lambda$ and the function $\boldsymbol{v}^\xi := \boldsymbol{D} \cdot \kappa \xi \in L^2(\Gamma)$. Indeed, given the formula (3.28) for the projected remainder and due to the assumption (A2), we can now exploit the following optimal estimate from [57, 58]: For all $0 < \gamma \le 1$, there exists a positive constant c such that

$$\int_0^\gamma d(E_\lambda \boldsymbol{v}^\xi, \boldsymbol{v}^\xi) \le c\gamma^{d/2+1}.$$

More precisely, we split the integral (3.28) into three parts, the first ranging from 0 to t^{-1}, the second from t^{-1} to 1 and the third from 1 to ∞, respectively when $t > 1$. For the latter we have the trivial estimate

$$\int_1^\infty d(E_\lambda \boldsymbol{v}^\xi, \boldsymbol{v}^\xi) \le \int_0^\infty d(E_\lambda \boldsymbol{v}^\xi, \boldsymbol{v}^\xi) = \mathbb{M}\{(\boldsymbol{v}^\xi)^2\}.$$

The first part is bounded by a positive constant as well, namely by

$$
\begin{aligned}
\int_0^{t^{-1}} t\lambda^{-1} \, d(E_\lambda v^\xi, v^\xi) &= t \int_0^{t^{-1}} \int_\lambda^\infty \alpha^{-2} \, d\alpha \, d(E_\lambda v^\xi, v^\xi) \\
&= t \int_0^\infty \alpha^{-2} \int_0^{\alpha \wedge t^{-1}} d(E_\lambda v^\xi, v^\xi) \\
&\leq ct \int_0^\infty \alpha^{-2} (\alpha \wedge t^{-1})^{d/2+1} \, d\alpha.
\end{aligned}
$$

Similarly, the second part can be estimated by

$$
\begin{aligned}
\int_{t^{-1}}^1 \lambda^{-2} \, d(E_\lambda v^\xi, v^\xi) &= 2 \int_{t^{-1}}^1 \int_\lambda^\infty \alpha^{-3} \, d\alpha \, d(E_\lambda v^\xi, v^\xi) \\
&\leq 2c \int_{t^{-1}}^\infty \alpha^{-3}(\alpha \wedge 1)^{d/2+1} \, d\alpha,
\end{aligned}
$$

which diverges logarithmically for $d = 2$ and is bounded by a positive constant for $d = 3$. Therefore, combining these computations with the identities (3.25) and (3.28) the estimate (3.23) follows.

It remains to prove the spectral representation (3.28). In order to take advantage of the spectral theorem, we would like to express the projected remainder in the form $\mathbb{M}\psi_1(\mathcal{L})v^\xi \psi_2(\mathcal{L})v^\xi$ for some bounded continuous functions ψ_1 and ψ_2. However, for this we would need the components of η to be stationary, which is not the case. Furthermore, we would want to use $\psi_1(x) = x^{-1}$ which is unbounded at zero.

Inspired by the computations in [132], we can find a remedy for both of these obstructions; namely we consider the function

$$
R_t^{\omega,\delta} := -\eta_\delta(X_t^\omega, \omega) + \eta_\delta(0, \omega),
$$

where η_δ is defined in analogy to η with the difference that it corresponds to a different auxiliary problem, modified by a zero order term. This modified auxiliary problem reads as follows: Find $\eta_\delta^\xi(\cdot, \omega) \in C(\mathbb{R}^d) \cap H_{\mathrm{loc}}^1(\mathbb{R}^d)$, $\delta > 0$ such that for $\mathcal{P}$-a.e. $\omega \in \Gamma$, the random field $\{\eta_\delta^\xi(x, \omega), (x, \omega) \in \mathbb{R}^d \times \Gamma\}$ is stationary with respect to $\mathcal{P}$ with $\mathbb{M}\eta_\delta^\xi(x, \cdot) = 0$ for every $x \in \mathbb{R}^d$ and satisfies

$$
\delta \eta_\delta^\xi(\cdot, \omega) - \nabla \cdot \kappa(\cdot, \omega)(\xi + \nabla \eta_\delta^\xi(\cdot, \omega)) = 0 \quad \text{in } \mathbb{R}^d
$$

for $\mathcal{P}$-a.e. $\omega \in \Gamma$. We refer to [132] for the proof of existence and uniqueness of η_δ^ξ, $\delta > 0$. Note that $\eta_\delta^\xi(0, \omega) \neq 0$ in general so that we have for the projected modified remainder the expression

$$
\overline{\mathbb{E}}(R_t^{\cdot,\delta} \cdot \xi)^2 = \overline{\mathbb{E}}(\eta_\delta^\xi(X_t))^2 - 2\overline{\mathbb{E}}\eta_\delta^\xi(X_t, \cdot)\eta_\delta^\xi(0, \cdot) + \overline{\mathbb{E}}(\eta_\delta^\xi(0, \cdot))^2. \tag{3.29}
$$

The equivalent formulation on $L^2(\Gamma)$ of the modified auxiliary problem for the direction $\xi \in \mathbb{R}^d$ reads as follows: Find the unique solution $\eta_\delta^\xi \in L^2(\Gamma)$ of the elliptic equation

$$\delta\eta_\delta^\xi - \boldsymbol{D} \cdot \boldsymbol{\kappa}(\xi + \boldsymbol{D}\eta_\delta^\xi) = 0 \quad \text{in } \Gamma.$$

In particular, the function $\boldsymbol{v}^\xi$ was chosen such that $\boldsymbol{v}^\xi = (\delta + \boldsymbol{\mathcal{L}})\eta_\delta^\xi = \boldsymbol{\mathcal{L}}\eta^\xi$. We will now only need to show that the modified remainder for fixed δ and fixed t can be written in the form $\mathbb{M}\psi_1(\boldsymbol{\mathcal{L}})\boldsymbol{v}^\xi\psi_2(\boldsymbol{\mathcal{L}})\boldsymbol{v}^\xi$, where $\psi_1(x) = 2(x+\delta)^{-1}$ and $\psi_2(x) = 2e^{-tx}(x+\delta)^{-1}$. In fact, the introduction of the zero order perturbation removes the singularity coming from the term x^{-1}.

The first term on the right-hand side of (3.29) may equivalently be written as

$$\overline{\mathbb{E}}(\eta_\delta^\xi(X_t^{\cdot}, \cdot))^2 = \overline{\mathbb{E}}(\eta_\delta^\xi(0, Y_t^{\cdot}))^2 = \mathbb{M}(\eta_\delta^\xi(0, \cdot))^2,$$

where we have used the stationarity of the environment as viewed from the particle with respect to $\overline{\mathbb{P}}$ and the stationarity of the random field $\{\eta_\delta^\xi(x, \omega), (x, \omega) \in \mathbb{R}^d \times \Gamma\}$ with respect to $\mathcal{P}$, respectively. Therefore, treating the second term on the right-hand side of (3.29) we obtain

$$\overline{\mathbb{E}}(R_t^{\cdot,\delta} \cdot \xi)^2 = 2\mathbb{M}(\eta_\delta^\xi(0, \cdot))^2 - 2\mathbb{M}\eta_\delta^\xi(0, \cdot)T_t^{\cdot}\eta_\delta^\xi(0, \cdot), \qquad (3.30)$$

where $\{T_t^\omega, t \geq 0\}$ denotes the strongly continuous semigroup on $L^2(\mathbb{R}^d)$ associated with $(\mathcal{E}^{\omega,1}, H^1(\mathbb{R}^d))$ which satisfies

$$\mathbb{E}_0^\omega\eta_\delta^\xi(X_t^\omega, \omega) = T_t^\omega\eta_\delta^\xi(0, \omega).$$

Going back to (3.30), respectively the corresponding identity on $L^2(\Gamma)$, we have thus for the first term on the right-hand side

$$2\mathbb{M}(\eta_\delta^\xi)^2 = 2\mathbb{M}(\delta + \boldsymbol{\mathcal{L}})^{-1}\boldsymbol{v}^\xi(\delta + \boldsymbol{\mathcal{L}})^{-1}\boldsymbol{v}^\xi = 2\int_0^\infty (\delta + \lambda)^{-2}\,\mathrm{d}(E_\lambda\boldsymbol{v}^\xi, \boldsymbol{v}^\xi),$$

whereas the second term may be written in the form

$$2\mathbb{M}(\delta + \boldsymbol{\mathcal{L}})^{-1}\boldsymbol{v}^\xi e^{-t\boldsymbol{\mathcal{L}}}(\delta + \boldsymbol{\mathcal{L}})^{-1}\boldsymbol{v}^\xi = 2\int_0^\infty (\delta + \lambda)^{-2}e^{-t\lambda}\,\mathrm{d}(E_\lambda\boldsymbol{v}^\xi, \boldsymbol{v}^\xi).$$

Altogether we have obtained

$$\overline{\mathbb{E}}(R_t^{\cdot,\delta} \cdot \xi)^2 = 2\int_0^\infty (1 - e^{-\lambda t})(\delta + \lambda)^{-2}\,\mathrm{d}(E_\lambda\boldsymbol{v}^\xi, \boldsymbol{v}^\xi)$$

and the right-hand side of this equality converges as $\delta \to 0$ so that the spectral representation (3.28) holds. $\qquad\square$

3.4 Numerical experiments

In this section, we provide numerical experiments to assess the capabilities of the proposed continuum micro-scale Monte Carlo method. The program code for the benchmark tests is implemented in CUDA C using single-precision floating point arithmetic. Pseudo random numbers were generated with an implementation of L'Ecuyer's parallel MRG32k3a random number generator, cf. [99], from the cuRAND library. The numerical experiments were performed on a standard GeForce GTX 680 graphics card with 1.536 computing kernels and 2 GB of RAM. The program code for the generation of the log-normal random fields is implemented in MATLAB R2013A.

3.4.1 Benchmark tests

All of the reported approximations were obtained using $M_1 = 5 \times 10^6$, respectively $M_1 = 10^6$, realizations of the random medium. Although the variance of the estimator (3.21) clearly depends on the choice of the number M_2 of paths per realization of the random medium, we did not attempt to optimize this choice but rather set $M_2 = 1$ throughout this section. For the sake of readability, we will therefore suppress the dependence on M_1 and M_2. Note moreover, that we have chosen benchmark tests where the effective conductivity is isotropic, i.e., $\kappa^* = cI_2$, and that we will identify κ^* with the scalar constant c.

For each approximation $\hat{\kappa}_T^*$, an estimate of its *standard error* e^{std}, i.e., the empirical standard deviation divided by the square root of the sample size is provided. The corresponding 95% confidence intervals are then given by

$$[\hat{\kappa}_T^* - 1.96 \times e^{\text{std}}, \hat{\kappa}_T^* + 1.96 \times e^{\text{std}}].$$

Two-phase media with i.i.d. phases

It is well known that checkerboard-like digitized random media with independent identically distributed (i.i.d.) conductivity in each tile constitute a severe benchmark test for numerical homogenization schemes. We have considered a checkerboard consisting of $n \times n$ squares which have either conductivity κ_1 or κ_2 with probability ϕ_1 and ϕ_2, respectively. In the special case of equal volume fractions $\phi_1 = \phi_2 = 1/2$, the exact effective conductivity is known, namely $\kappa^* = \sqrt{\kappa_1 \kappa_2}$. For other volume fractions, there exist no known analytical expressions for the effective conductivity. However, due to the computational challenges of numerical homogenization, practitioners often avoid approximating κ^* but rather rely on available a priori bounds. For the case of two-phase media with i.i.d. phases, the so-called *Hashin-Shtrikman bounds* are presumably the most prominent ones, cf. [158]. These bounds are sharp, but their accuracy strongly depends on both the conductivity contrast and the volume fraction. In our first benchmark test, we have used the proposed continuum micro-scale Monte Carlo method to approximate the effective conductivity for different conductivity contrasts and different volume fractions in two and three space dimensions and compared them to their known exact values or to the Hashin-Shtrikman bounds.

A similar benchmark test using a finite element based Monte Carlo method with variance reduction via antithetic variables has been carried out by Costaouec, Le Bris and Legoll [37].

Log-normal media

In the second benchmark test, we have considered (digitized) realizations of an isotropic log-normal random field $\{Z(x,\omega), (x,\omega) \in D \times \Gamma\}$ with constant mean Z_0 in two and three space dimensions, that is,

$$Z(x,\omega) = \exp(G(x,\omega)),$$

where $\{G(x,\omega), (x,\omega) \in D \times \Gamma\}$ is a Gaussian random field whose covariance function depends only on the distance of its arguments. For the two-dimensional case, we simulated the corresponding random fields exactly by the method of *circulant embedding* introduced by Dietrich and Newsam [41]. See Appendix B for a brief description of this method. In the three-dimensional case, we used approximate random field simulation via the so-called *turning bands method*, cf. [119].

In the two-dimensional case, the effective conductivity is given by the geometric mean $\kappa^* = \exp(G_0)$, cf., e.g., [75], while in the three-dimensional case, there is no known analytical expression.

We have run tests for two different Gaussian random fields with covariance functions from the Matérn class, namely the squared exponential covariance function

$$\mathcal{C}_G^\infty(x,y) = \sigma^2 \exp\left(-\frac{|x-y|^2}{2\lambda^2}\right) \tag{3.31}$$

and the exponential covariance function

$$\mathcal{C}_G^{1/2}(x,y) = \sigma^2 \exp\left(-\frac{|x-y|}{\lambda}\right) \tag{3.32}$$

with parameters $G_0 = 0$, $\sigma = \sqrt{2}$, $\lambda = 0.02$. Two-dimensional realizations of such fields are plotted in Figure 3.5 and Figure 3.6, respectively. A similar

Figure 3.5: Realization of a log-normal random field with squared exponential covariance function.

Figure 3.6: Realization of a log-normal random field with exponential covariance function.

benchmark test using a finite volume based Monte Carlo method, where the random field is approximated via the *Karhunen-Loève expansion*, has been carried out in the monograph [94] by Kronsbein.

3.4.2 Results and discussion

Two-phase media with i.i.d. phases

For the particular case of a two-dimensional two-phase medium with i.i.d. phases and equal volume fractions $\phi_1 = \phi_2 = 1/2$, we first studied the bias due to the finite time horizon. To keep the Monte Carlo error reasonably small, we used $M_1 = 5 \times 10^6$ realizations of the random medium. Our findings for two different conductivity contrasts are plotted in Figure 3.7, respectively Figure 3.8. In the first case, $\kappa_1/\kappa_2 = 2$, the standard error was at most 6.3×10^{-4}, see also Table 3.1 below (note that the standard error increases slightly with T), whereas the combined error for the largest time horizon $T = 30$ was 7.0×10^{-4}.

In the second case, $\kappa_1/\kappa_2 = 20/3$, the standard error was at most 4.1×10^{-3}, see also Table 3.1 below, whereas the combined error for the largest time horizon $T = 30$ was 4.9×10^{-3}.

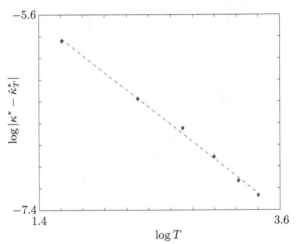

Figure 3.7: First benchmark test: 2D i.i.d. phases and equal volume fractions; dashed line is the least squares fit with slope -0.794; $\kappa_1/\kappa_2 = 2$.

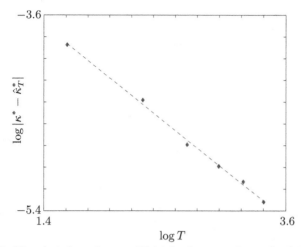

Figure 3.8: First benchmark test: 2D i.i.d. phases and equal volume fractions; dashed line is the least squares fit with slope -0.807; $\kappa_1/\kappa_2 = 20/3$.

Subsequently, we approximated the effective conductivity for different conductivity contrasts. The results for time horizon $T = 30$ are collected in Table 3.1.

These results are very encouraging with comparably high accuracy even for large conductivity contrasts. In fact, the approximation in the last row of Table 3.1 is slightly better than the already rather accurate approximation 10.063 which was obtained in [159]. This is indeed no surprise as the proposed method is, in some sense, an advancement of the method in [159] which uses mean exit times rather than exact simulation. Notice finally, that for all of the experiments, the exact value lies within the 95% confidence interval.

κ_1	κ_2	$\hat{\kappa}_T^*$	e^{std}
1	1	1.000	4.4×10^{-4}
2	1	1.415	6.3×10^{-4}
20	3	7.739	4.1×10^{-3}
100	1	9.989	5.7×10^{-3}

Table 3.1: First benchmark test: 2D i.i.d. phases and equal volume fractions.

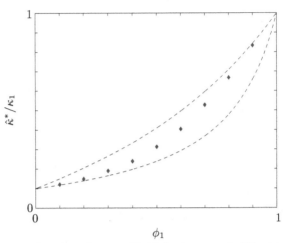

Figure 3.9: First benchmark test: 2D i.i.d. phases and different volume
fractions; scaled approximation and Hashin-Shtrikman bounds;
$\kappa_1/\kappa_2 = 10$.

Moreover, we computed approximations for different volume fractions in
two and three space dimensions. The results for the two-dimensional case
and conductivity contrast $\kappa_1/\kappa_2 = 10$, respectively $\kappa_1/\kappa_2 = 100$ are plotted
in Figure 3.9 and Figure 3.10 together with the corresponding Hashin-
Shtrikman bounds. Again, it can be observed that our approximations
coincide fairly well with the values reported in [159] that are believed to be
rather accurate. For the three dimensional case, which is plotted in Figure
3.11, we could not locate any reference values in the literature, therefore we
confine ourselves with a comparison to the theoretical bounds from [158].

At the so-called *site percolation threshold*, cf. [158], $\phi_c \approx 0.5927$ if $d = 2$
and $\phi_c \approx 0.3116$ if $d = 3$, respectively, large clusters of constant conductivity
can be found and long-range connectivity first appears. It can be observed
from the plots in Figure 3.9, Figure 3.10 and Figure 3.11 that below this
percolation threshold, the scaled approximations lie closer to the lower bound,
whereas they lie closer to the upper bound above the threshold. Such a
behavior is well known and has been reported, e.g., in [157, 159].

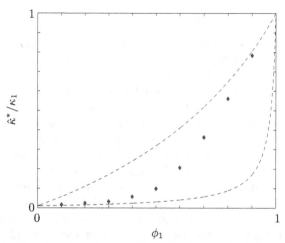

Figure 3.10: First benchmark test: 2D i.i.d. phases and different volume fractions; scaled approximation and Hashin-Shtrikman bounds; $\kappa_1/\kappa_2 = 100$.

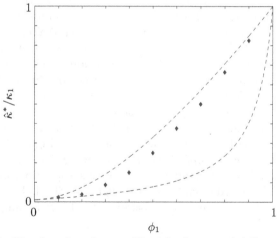

Figure 3.11: First benchmark test: 3D i.i.d. phases and different volume fractions; scaled approximation and theoretical bounds; $\kappa_1/\kappa_2 = 100$.

Log-normal media

For the particular case of a two-dimensional log-normal random medium with covariance function (3.31) and (3.32), respectively, we have approximated the effective conductivity. The results for time horizon $T = 10$ and $M_1 = 10^6$ realizations of the medium are collected in Table 3.2. In both experiments the exact value lies within the 95% confidence interval. Moreover, we would like to emphasize that for the particular case of the squared exponential covariance function (3.31), the accuracy of our method is of the same order of magnitude as the accuracy reported in [94]. In fact, it seems that the accuracy of the latter method is strongly dependent on both the regularity of the covariance function and the spatial dimension. More precisely, the squared exponential covariance function is analytic, implying that the corresponding Karhunen-Loève expansion converges rapidly; on top of that, the realizations of the random field are smooth. On the other hand, working with the exponential covariance function (3.32) poses a major challenge for such methods because of the slow convergence of the Karhunen-Loève expansion as well as low spatial regularity of the random field realizations. It can be observed from the result in the second row of Table 3.2 that our method does not suffer from these drawbacks. In contrast to finite volume or finite element based methods, the use of the Karhunen-Loève expansion is not mandatory here. Indeed, one should rather rely on efficient and *exact* random field simulation via the circulant embedding technique. Moreover, the convergence rate of our method does neither depend on the spatial regularity of the underlying random field nor on the spatial dimension.

For a three-dimensional log-normal random medium with covariance function (3.31), we have plotted the approximations for different standard deviations σ of the underlying Gaussian random field in Figure 3.12 together with the so-called *De Wit approximation*, cf. [40]. Somewhat surprisingly, the approximations differ significantly from this approximation, even for relatively small standard deviations. Although a similar behavior has been described in [81] for a finite element based method, our findings should be

covariance	$\hat{\kappa}_T^*$	e^{std}
(3.31)	1.012	7.6×10^{-3}
(3.32)	0.983	9.1×10^{-3}

Table 3.2: Second benchmark test: 2D log-normal random field.

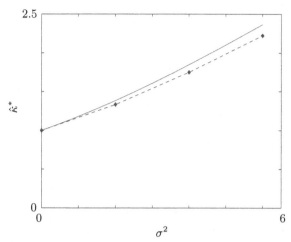

Figure 3.12: Second benchmark test: 3D log-normal random field with squared exponential covariance function; approximate effective conductivity and De Wit approximation (solid line).

considered an experimental result. In fact, in contrast to the two-dimensional case, we did not simulate the underlying random fields exactly but by the turning bands method whose error is somewhat difficult to assess.

Conclusion

To sum up, it can be stated that we have introduced and analyzed a relatively simple continuum micro-scale Monte Carlo method which is competitive and in some cases, such as three-dimensional problems or problems with rough conductivities, even is expected to outperform state-of-the-art homogenization methods based on finite element or finite volume discretization. Most importantly, due to its inherent parallelism, the proposed method is perfectly suited for modern multi- and many-core processor architectures. Future work will concentrate on variance reduction techniques in order to reduce the Monte Carlo error.

4 Statistical inversion

In this chapter, we connect the dots, that is, we turn to the inverse anomaly detection problem. More precisely, the numerical homogenization via the continuum micro-scale Monte Carlo method developed in the previous chapter serves as the basis for a two-stage numerical method in the framework of Bayesian inverse problems. The novelty of this method lies in the introduction of an enhanced error model accounting for the approximation errors that result from reducing the full forward model to a homogenized one. We proceed along the lines of the recent article [148] by the author: In the first stage, a MAP estimate for the reduced forward model equipped with the enhanced error model is computed. Then, in the second stage, a bootstrap prior based on the first stage results is defined and the resulting posterior distribution is sampled via Markov chain Monte Carlo. We provide the theoretical foundation of the proposed method, discuss different aspects of a numerical implementation and present numerical experiments to support our findings.

We start in Section 4.1 by recalling the basic ideas of statistical inversion in the Bayesian framework. In Section 4.2, we introduce the new error model which yields the two-stage method described subsequently in Section 4.3. Then, in Section 4.4, we provide numerical experiments to illustrate the feasibility of the proposed method. In addition, we demonstrate the advantage of the new error model, which accounts for the random homogenization error, over the conventional one.

4.1 The Bayesian framework

In the Bayesian framework, inverse problems are interpreted as problems of statistical inference, that is, all unknown quantities are modeled as random variables, cf., e.g. [78, 79, 92], and the objective is to extract information about some of these random variables based on knowledge of the measurements, the model and information which is available prior to the measurement. That is, the randomness reflects our uncertainty or lack of information about these things. In contrast to the traditional regularization methods, cf. [44], the solution to the statistical inverse problem is the joint posterior probability distribution of the sought parameters conditioned on the measurements rather than a single estimate.

Let us give a brief summary of the basic ideas, assuming that the variables are finite-dimensional random vectors. As all computational algorithms work on such finite-dimensional approximations, we restrict ourselves here to this case whilst for the infinite-dimensional function space viewpoint on the subject, we refer the reader to Stuart [154]. Throughout this chapter, we use bold capital letters for random vectors and bold lower case letters for their particular realizations.

Let $(\mathbf{X}, \mathbf{E}) \in \mathbb{R}^{M+M'}$ and $\mathbf{Y} \in \mathbb{R}^{M'}$ be random vectors, where the latter vector represents the quantities that are observable through measurements and the first vector consists of quantities that can not be directly measured. More precisely, $\mathbf{X}$ represents those unknowns we are primarily interested in, while $\mathbf{E}$ characterizes both, the modeling error and measurement error and we refer to this combined error term simply as *measurement noise*. For simplicity of the presentation, we use the common assumption that the noise is additive. More precisely, we assume that these variables are related via the *forward model*

$$\mathbf{Y} = F(\mathbf{X}) + \mathbf{E},$$

where $F : \mathbb{R}^M \to \mathbb{R}^{M'}$ denotes the *forward map*, i.e., a known deterministic function representing a model of the measurement. Moreover, we assume throughout this chapter, that the distributions of all random vectors are absolutely continuous with respect to the Lebesgue measure. The probability density of $\mathbf{Y}$ conditioned on $\mathbf{X} = \mathbf{x}$ and $\mathbf{E} = \mathbf{e}$ is given by

$$\pi(\mathbf{y}|\mathbf{x}, \mathbf{e}) = \delta(\mathbf{y} - (F(\mathbf{x}) + \mathbf{e})),$$

where δ denotes the Dirac distribution in $\mathbb{R}^{M'}$. The joint probability distribution of the vectors $(\mathbf{X}, \mathbf{E})$ and $\mathbf{Y}$ can be written as

$$\pi(\mathbf{x}, \mathbf{e}, \mathbf{y}) = \delta(\mathbf{y} - (F(\mathbf{x}) + \mathbf{e}))\pi(\mathbf{x}, \mathbf{e})$$

and marginalizing the noise we obtain

$$\pi(\mathbf{x}, \mathbf{y}) = \int_{\mathbb{R}^{M'}} \delta(\mathbf{y} - (F(\mathbf{x}) + \mathbf{e}))\pi(\mathbf{x}, \mathbf{e})\, d\mathbf{e}.$$

If we assume additionally that the variables $\mathbf{X}$ and $\mathbf{E}$ are independent, we may write

$$\pi(\mathbf{x}, \mathbf{e}) = \pi_{\text{prior}}(\mathbf{x})\pi_{\text{noise}}(\mathbf{e}),$$

where the densities on the right-hand side are called *prior densities*. Moreover, integration with respect to **e** may be carried out explicitly in this case which yields

$$\pi(\mathbf{x}, \mathbf{y}) = \pi_{\text{prior}}(\mathbf{x})\pi_{\text{noise}}(\mathbf{y} - F(\mathbf{x})).$$

The *solution* to the statistical inverse problem is defined as the *posterior distribution* of **X** conditioned on the measurement **y**, i.e., the conditional probability $\pi(\mathbf{x}|\mathbf{y})$ given by Bayes' formula

$$\pi(\mathbf{x}|\mathbf{y}) = \frac{\pi(\mathbf{x}, \mathbf{y})}{\pi(\mathbf{y})}. \tag{4.1}$$

Loosely speaking, in the Bayesian framework an ill-posed inverse problem is regularized by restating it as a well-posed extension in a larger space of probability distributions in the following sense: Let $\mu_{\mathbf{y}}$ denote the *posterior measure* of **x** corresponding to the measured data **y** which is related to the *prior measure* μ_{prior} through the Radon-Nikodym derivative

$$\frac{\mathrm{d}\mu_{\mathbf{y}}}{\mathrm{d}\mu_{\text{prior}}}(\mathbf{x}) \propto \pi_{\text{noise}}(\mathbf{y} - F(\mathbf{x})).$$

Let $\mathbf{E} \sim \mathcal{N}(0, \mathcal{C})$, then it can be shown under quite general assumptions on both, the forward model and the prior, that the posterior measure $\mu_{\mathbf{y}}$ is Lipschitz-continuous in the measurement data **y** with respect to the Hellinger distance. More precisely, if $\mu_{\mathbf{y}}$ and $\mu_{\mathbf{y}'}$ are two posterior measures corresponding to measurement data **y** and **y**′, respectively, satisfying

$$\max\{|\mathcal{C}^{-1/2}\mathbf{y}|, |\mathcal{C}^{-1/2}\mathbf{y}'|\} < r,$$

then there exists a positive constant $c = c(r)$ such that

$$d_{\text{Hell}}(\mu_{\mathbf{y}}, \mu_{\mathbf{y}'}) \leq c|\mathcal{C}^{-1/2}(\mathbf{y} - \mathbf{y}')|, \tag{4.2}$$

where the Hellinger distance of the measures $\mu_{\mathbf{y}}$ and $\mu_{\mathbf{y}'}$ is given by

$$d_{\text{Hell}}(\mu_{\mathbf{y}}, \mu_{\mathbf{y}'}) := \left(\frac{1}{2}\int \left(\sqrt{\frac{\mathrm{d}\mu_{\mathbf{y}}}{\mathrm{d}\mu_{\text{prior}}}} - \sqrt{\frac{\mathrm{d}\mu_{\mathbf{y}'}}{\mathrm{d}\mu_{\text{prior}}}}\right)^2 \mathrm{d}\mu_{\text{prior}}\right)^{1/2},$$

cf. [154].

4.1.1 Estimates from the posterior distribution

Given the solution to the statistical inverse problem, one often needs to compute point estimates and credibility intervals for these estimates.

A commonly used point estimate is the (possibly non-unique) *maximum a posteriori* (MAP) estimate

$$\mathbf{x}_{\mathrm{MAP}} = \arg\max_{\mathbf{x} \in \mathbb{R}^M} \pi(\mathbf{x}|\mathbf{y}), \qquad (4.3)$$

whose computation leads to an optimization problem. Under an additive independent Gaussian noise model and a Gibbs type prior distribution

$$\pi_{\mathrm{prior}}(\mathbf{x}) \propto e^{-\alpha \Psi(\mathbf{x})},$$

where Ψ is a *regularizing functional* and α a positive *regularization parameter*, the posterior density is of the form

$$\pi(\mathbf{x}|\mathbf{y}) \propto \exp\left(-\frac{1}{2} \|\mathbf{y} - F(\mathbf{x}))\|_{C^{-1}}^2 - \alpha \Psi(\mathbf{x}) \right),$$

where for vectors $\mathbf{v} \in \mathbb{R}^{M'}$ and matrices $A \in \mathbb{R}^{M' \times M'}$ we define $\|\mathbf{v}\|_A :=$ $\mathbf{v}^T A \mathbf{v}$. In this case, the MAP estimate

$$\mathbf{x}_{\mathrm{MAP}} = \arg\min_{\mathbf{x} \in \mathbb{R}^M} \left\{ \frac{1}{2} \|\mathbf{y} - F(\mathbf{x}))\|_{C^{-1}}^2 + \alpha \Psi(\mathbf{x}) \right\}$$

coincides with the classical Tikhonov regularization, cf. [44].

Remark 4.1. Note that the well-known *maximum likelihood* estimate

$$\mathbf{x}_{\mathrm{ML}} = \arg\max_{\mathbf{x} \in \mathbb{R}^M} \pi(\mathbf{y}|\mathbf{x})$$

is not stable in the case of ill-posed problems as it corresponds to the solution of the problem without regularization due to available prior information.

Another commonly used estimate is the *conditional mean* estimate

$$\mathbf{x}_{\mathrm{CM}} = \int_{\mathbb{R}^M} \mathbf{x}\pi(\mathbf{x}|\mathbf{y}) \, d\mathbf{x}, \qquad (4.4)$$

whose computation corresponds to an integration problem.

Assume now that $\hat{\mathbf{x}} \in \mathbb{R}^M$ is an estimate obtained from the posterior distribution. To assess the reliability of this estimate, it is useful to calculate credibility intervals from the *marginal posterior probability density* of the components $\mathbf{X}_k$, $k = 1, ..., M$, given by

$$\pi(\mathbf{x}_k|\mathbf{y}) \propto \int_{\mathbb{R}^{M-1}} \pi(\mathbf{x}|\mathbf{y}) \, \mathrm{d}(\mathbf{x}, ..., \mathbf{x}_{k-1}, \mathbf{x}_{k+1}, ...\mathbf{x}_M)^T \qquad (4.5)$$

as well as the *conditional correlation matrix*

$$\int_{\mathbb{R}^M} (\mathbf{x} - \hat{\mathbf{x}})(\mathbf{x} - \hat{\mathbf{x}})^T \pi(\mathbf{x}|\mathbf{y}) \, \mathrm{d}\mathbf{x}. \qquad (4.6)$$

4.2 Error modeling

We turn to the setting of the stochastic forward problem described in Chapter 1 and for simplicity of the presentation, we assume that $d = 2$. Let the conductivity random field satisfy assumptions (A1) and (A2) from Subsection 1.3.1 and assume, on top of that, that the effective conductivity obtained from Theorem 3.2 is isotropic. A sufficient condition for this is that κ is Q-*invariant* for a given $Q \in SO(2)\backslash\{\pm I_2\}$, i.e., there exists an automorphism ζ on Γ such that

$$\kappa(x, \zeta(\omega)) = Q\kappa(Q^T x, \omega)Q^T$$

for a.e. $x \in \mathbb{R}^2$ and $\mathcal{P}$-a.e. $\omega \in \Gamma$, cf. [4, Proposition 4].

4.2.1 Homogenization error

As we aim to study the inverse problem of reconstructing the region Σ in the Bayesian framework, let us assume that it is a random quantity which is parameterized by the following *anomaly prior* with density π^0_{prior}: Assume that the domain Σ is a disc of radius r centered at $(c_1, c_2)^T$ and that each of those parameters is a uniformly distributed random variable on some interval. A realization $\mathbf{x}_a = (c_1, c_2, r)^T$ drawn from the anomaly prior is said to be *admissible* if $\Sigma(\mathbf{x}_a)$ belongs to D and does not touch the boundary of the domain D. That is, there is a one-to-one correspondence between the set of admissible anomalies and the set $\mathcal{A}$ of admissible parameter vectors.

Clearly, the homogenization result of Theorem 3.2 for the finite-dimensional linear voltage-to-current map may be equivalently formulated in terms of its inverse, the current-to-voltage map which is used more commonly in practical applications. Therefore, we assume that we can measure random voltage patterns $U^{(k)}(\varepsilon, \omega, \mathbf{x}_a)$ corresponding to the linearly independent deterministic current patterns $J^{(k)}$, $k = 1, ..., N - 1$, and we denote the con-catenated vector containing all these measurement patterns by $\mathbf{U}(\kappa_\varepsilon|_{D\backslash\overline{\Sigma}})$.

Analogously we denote by $\overline{\mathbf{U}}(\kappa^*|_{D\setminus\overline{\Sigma}})$ the random voltage vector corresponding to the isotropic effective background conductivity κ^*, perturbed by a realization $\mathbf{x}_a$ of the anomaly. Note that we suppress the dependence on ω and $\mathbf{x}_a$ whenever this causes no confusion. With regard to Theorem 3.2, it seems natural to consider the random *homogenization error*

$$\mathbf{U}(\kappa_\varepsilon|_{D\setminus\overline{\Sigma}}) - \overline{\mathbf{U}}(\kappa^*|_{D\setminus\overline{\Sigma}}) \tag{4.7}$$

for the stochastic forward problem (1.16), (1.17).

Our approach is inspired by the *approximation error approach* in statistical inverse problems, cf. [78, 79]. Indeed, we are going to utilize the homogenization error (4.7) in conjunction with an approximation error model to account for the modeling error introduced by neglecting the random microstructure of the medium and using the homogenized forward model (3.7), (3.8) instead. Consider measurements of the form

$$\mathbf{y}_\varepsilon = \mathbf{u}(\kappa_\varepsilon|_{D\setminus\overline{\Sigma}}) + \mathbf{e},$$

where the vector $\mathbf{e}$ is a particular realization of the *measurement noise process* $\mathbf{E}$, which we assume to be a centered Gaussian random vector with a known covariance matrix $\mathcal{C}$. Then we may use the homogenization error (4.7) to write equivalently

$$\begin{aligned}
\mathbf{Y}_\varepsilon &= \overline{\mathbf{U}}(\kappa^*|_{D\setminus\overline{\Sigma}}) + (\mathbf{U}(\kappa_\varepsilon|_{D\setminus\overline{\Sigma}}) - \overline{\mathbf{U}}(\kappa^*|_{D\setminus\overline{\Sigma}})) + \mathbf{E} \\
&= \overline{\mathbf{U}}(\kappa^*|_{D\setminus\overline{\Sigma}}) + \mathbf{E}_\varepsilon,
\end{aligned} \tag{4.8}$$

where $\mathbf{E}_\varepsilon$ denotes the combined noise term accounting for both measurement and homogenization error. Following [78], we introduce an *enhanced error model*, i.e., an independent Gaussian approximation $\widehat{\mathbf{E}}_\varepsilon$ of this noise term with mean m_ε and covariance matrix $\mathcal{C}_\varepsilon$ given by

$$\begin{aligned}
m_\varepsilon &= \int_{\mathcal{A}} \int_\Omega (\mathbf{U}(\kappa_\varepsilon|_{D\setminus\overline{\Sigma}}) - \overline{\mathbf{U}}(\kappa^*|_{D\setminus\overline{\Sigma}}))\pi^0_{\text{prior}}(\mathbf{x}_a)\,\mathrm{d}\mathcal{P}(\omega)\,\mathrm{d}\mathbf{x}_a, \\
\mathcal{C}_\varepsilon &= \int_{\mathcal{A}} \int_\Omega (\mathbf{U}(\kappa_\varepsilon|_{D\setminus\overline{\Sigma}}) - \overline{\mathbf{U}}(\kappa^*|_{D\setminus\overline{\Sigma}}))(\mathbf{U}(\kappa_\varepsilon|_{D\setminus\overline{\Sigma}}) - \overline{\mathbf{U}}(\kappa^*|_{D\setminus\overline{\Sigma}}))^T \\
&\quad \times \pi^0_{\text{prior}}(\mathbf{x}_a)\,\mathrm{d}\mathcal{P}(\omega)\,\mathrm{d}\mathbf{x}_a + \mathcal{C}.
\end{aligned} \tag{4.9}$$

Remark 4.2. With the notable exception of Schrödinger operators with random potentials and random divergence form operators in 1D, cf. [49, 130], there are in general no known closed form expressions for the homogenization errors. Hence their statistics must be approximated via Monte Carlo sampling.

Let us now consider a discretized forward problem by identifying the deviations of the true conductivity (including the anomaly) from the effective conductivity κ^* with the coefficient vector $\mathbf{x} \in \mathbb{R}^M$, i.e., we approximate the conductivity in D by

$$\kappa^* + \sum_{m=1}^{M} \mathbf{x}_m [\mathcal{T}_m^h](x), \qquad (4.10)$$

where the basis functions $[\mathcal{T}_m^h]$ are the characteristic functions of the underlying finite element triangulation with discretization parameter h and M is the number of elements, see Appendix C for details on the discretization of the forward problem corresponding to the complete electrode model. The unknown coefficient vector $\mathbf{x}$ is modeled as an M-variate random vector and $F_{h,D}(\cdot; \kappa^*)$ denotes the discrete forward map, which maps realizations of $\mathbf{x}$ to the approximate voltage vector obtained by the finite element method for the discretized conductivity (4.10). The discrete *accurate measurement model* based on the enhanced error model (4.8), (4.9) is

$$\mathbf{Y} = F_{h,D}(\mathbf{x}; \kappa^*) + \widehat{\mathbf{E}}_\varepsilon, \qquad (4.11)$$

where the discretization parameter h is chosen sufficiently small such that the discretization error related to the finite element approximation is dominated by the combined noise term accounting for both, measurement noise and homogenization error. In order to regularize the problem, we consider Gibbs type prior distributions of the form

$$\pi_{\text{prior}}(\mathbf{x}) \propto \pi^+(\mathbf{x}) e^{-\alpha \Psi(\mathbf{x})},$$

where π^+ is a *positivity prior* and the regularizing functional Ψ corresponds to a Gaussian *smoothness prior*, i.e., $\Psi(\mathbf{x}) := \mathbf{x}^T A \mathbf{x}$. As in [139], the matrix $A \in \mathbb{R}^{M \times M}$ is a finite element stiffness matrix for the Laplace equation with homogeneous Dirichlet boundary condition. We refer to this article for a detailed description of the construction of such smoothness priors.

Obviously, the better prior information we have, the better the estimates we can obtain. If available, we could therefore include even more precise prior information about the correlation function of the underlying random field. In the recent paper [144], Roininen, Piiroinen and Lehtinen have considered lattice approximation of certain Matérn priors; see also the recent work [143] by Roininen, Huttunen and Lasanan for a numerical method using Matérn priors for statistical inversion in EIT.

Remark 4.3. The particular choice of the boundary condition for the smoothness prior is based on our numerical experiments. To be precise, we have observed that the presence of the Dirichlet boundary condition seems to reduce artefacts in the vicinity of the electrodes caused by the rapidly oscillating background conductivity occurring in the Robin-type boundary condition (1.17). Moreover, it seems to increase the sensitivity with respect to anomalies that are located further away from the electrodes.

Remark 4.4. Note that the realizations drawn from the prior density π_{prior} are quite different from the conductivities to be imaged which present both, sharp jumps and an anomaly where the conductivity is infinite. On the other hand, Gaussian smoothness priors are appealing in practice due to their differentiability which is necessary if one wants to apply Newton-type methods. Moreover, it has been shown in [6, 100] that the approximation error approach is to some extent tolerant when it comes to misspecification of the prior. We would nevertheless like to point out, that in the Bayesian framework it would be possible to use priors that allow moderate discontinuities of the conductivity, however, leading to non-differentiable prior functionals, cf. [77].

The solution to the statistical inverse problem based on the enhanced error model (4.8), (4.9) is the posterior density

$$\pi(\mathbf{x}|\mathbf{y}) \propto \pi^+(\mathbf{x}) \exp\left(-\frac{1}{2}\|\mathbf{y} - F_{h,D}(\mathbf{x}; \kappa^*) - m_\varepsilon\|^2_{\mathcal{C}_\varepsilon^{-1}} - \alpha\Psi(\mathbf{x})\right).$$

4.2.2 Approximation errors

Minimizing discretization and modeling errors by using the high-resolution forward map $F_{h,D}(\cdot; \kappa^*)$, where the whole model domain D is discretized, may be impractical because of computational limitations. Indeed, it is often necessary to use a *reduced forward model* $F_{h_c,D_c}(\cdot; \kappa^*)$ with truncated computational domain $D_c \subset D$ and coarser discretization corresponding to the discretization parameter $h_c > h$. Assume that there is a linear model reduction map

$$P : \mathbb{R}^M \to \mathbb{R}^{M_c}, \quad \mathbf{x} \mapsto \mathbf{x}_c$$

so that the accurate measurement model can be written as

$$\begin{aligned}\mathbf{Y} &= F_{h,D}(\mathbf{x}; \kappa^*) + (F_{h_c,D_c}(P\mathbf{x}; \kappa^*) - F_{h_c,D_c}(P\mathbf{x}; \kappa^*)) + \widehat{\mathbf{E}}_\varepsilon \\ &= F_{h_c,D_c}(\mathbf{x}_c; \kappa^*) + \mathbf{E}_{\varepsilon,c},\end{aligned} \tag{4.12}$$

where the random variable $\mathbf{E}_{\varepsilon,c}$ is given by the sum of the discrepancy between the accurate and the reduced model and the combined noise term accounting for measurement and homogenization error. Analogously to the procedure in the previous subsection, we approximate this noise term by an enhanced error model, i.e., an independent Gaussian random variable $\widehat{\mathbf{E}}_{\varepsilon,c}$ with mean $m_{\varepsilon,c}$ and covariance matrix $\mathcal{C}_{\varepsilon,c}$. Hence, we obtain the *reduced* posterior density

$$\pi(\mathbf{x}_c|\mathbf{y}) \propto \pi^+(\mathbf{x}_c) \exp\left(-\frac{1}{2}||\mathbf{y} - F_{h_c,D_c}(\mathbf{x}_c; \kappa^*) - m_{\varepsilon,c}||^2_{\mathcal{C}^{-1}_{\varepsilon,c}} - \alpha\Psi(\mathbf{x}_c)\right). \tag{4.13}$$

4.3 Reconstruction method

We adopt the two-stage strategy proposed by Pursiainen [139]: In the first stage, a region of interest (ROI) is determined using the Gauss-Newton algorithm to approximate the MAP estimate (4.3) for the reduced posterior density (4.13). The final reconstruction is found in the second stage using MCMC sampling from a *bootstrap prior* $\pi^{\text{ROI}}_{\text{prior}}(\mathbf{x}_a)$. More precisely, the intervals for the uniformly distributed components of $\mathbf{x}_a$, which are encoded in $\pi^0_{\text{prior}}(\mathbf{x}_a)$, are adapted according to the information obtained from the ROI.

Remark 4.5. It is pointed out in [139] that the resulting reconstruction strategy is not fully Bayesian since the ROI is used as *a priori* information in the second stage.

4.3.1 First stage: MAP estimate

The first stage of the proposed method corresponds to the classical Tikhonov regularization, cf. [44], however, with a carefully chosen error model. To be precise, we use the Gauss-Newton method to minimize the functional

$$\Phi_\alpha(\mathbf{x}_c) := \frac{1}{2}||\mathbf{y} - F_{h_c,D_c}(\mathbf{x}_c; \kappa^*) - m_{\varepsilon,c}||^2_{\mathcal{C}^{-1}_{\varepsilon,c}} + \alpha\Psi(\mathbf{x}_c),$$

cf. Algorithm 5, respectively the Appendix of [77], where the computation of the Jacobian matrix $DF_{h_c,D_c}(\mathbf{x}_c; \kappa^*)$ is described. Once we have obtained an approximation $\hat{\mathbf{x}}_{c,\text{MAP}}$ of the MAP estimate for the reduced posterior density, we determine a ROI potentially containing the anomaly by

$$\mathcal{R} = \{\mathcal{T}^{h_c}_m : (\hat{\mathbf{x}}_{c,\text{MAP}})_m > t\}$$

for some thresholding parameter $t > 0$.

Algorithm 5 Gauss-Newton method

Require: initial guess $\mathbf{x}_c^{(0)}$, parameter $\beta \in (0, 1]$, number of steps n
    for $k = 0 : n - 1$ **do**

$$\nabla \Phi_\alpha(\mathbf{x}_c^{(k)}) \leftarrow DF_{h_c,D_c}(\mathbf{x}_c^{(k)}; \kappa^*)^T \mathcal{C}_{\varepsilon,c}^{-1}(\mathbf{y} - F_{h_c,D_c}(\mathbf{x}_c^{(k)}; \kappa^*)) + \alpha D\Psi(\mathbf{x}_c^{(k)})$$

$$D\nabla \Phi_\alpha(\mathbf{x}_c^{(k)}) \leftarrow DF_{h_c,D_c}(\mathbf{x}_c^{(k)}; \kappa^*)^T \mathcal{C}_{\varepsilon,c}^{-1} DF_{h_c,D_c}(\mathbf{x}_c^{(k)}; \kappa^*) + \alpha D^2\Psi(\mathbf{x}_c^{(k)})$$

$$\mathbf{x}_c^{(k+1)} \leftarrow \mathbf{x}_c^{(k)} - \beta(D\nabla \Phi_\alpha(\mathbf{x}_c^{(k)}))^{-1} \nabla \Phi_\alpha(\mathbf{x}_c^{(k)})$$

    end for
    return $\mathbf{x}_c^{(n)}$

4.3.2 Second stage: MCMC bootstrap

In the second stage we assume, as in the approximation of the homogenization error statistics, that the position and radius of the circular anomaly are distributed according to an anomaly prior. To be precise, the ROI obtained in the first stage yields (component-wise) bounds for $\mathbf{x}_a$ and we assume that the components of $\mathbf{x}_a$ are uniformly distributed within the corresponding intervals. Moreover, we assume that the conductivity is of the form $\kappa^*|_{D \setminus \overline{\Sigma}}$. The prior $\pi_{\mathrm{prior}}^{\mathrm{ROI}}(\mathbf{x}_a)$ incorporating these assumptions is called the second stage *bootstrap prior*. As in the first stage, we use a discrete measurement model

$$\mathbf{Y} = \widetilde{F}_{h_c,D_c}(\mathbf{x}_a; \kappa^*) + \widetilde{\mathbf{E}}_{\varepsilon,c},$$

where the discrete forward map $\widetilde{F}_{h_c,D_c}(\cdot; \kappa^*)$ maps realizations of $\mathbf{x}_a$ to the approximate voltage vector and the enhanced error model is defined as in the first stage. Hence, we obtain the *bootstrap posterior density*

$$\pi(\mathbf{x}_a|\mathbf{y}) \propto \pi_{\mathrm{prior}}^{\mathrm{ROI}}(\mathbf{x}_a) \exp\left(-\frac{1}{2}\|\mathbf{y} - \widetilde{F}_{h_c,D_c}(\mathbf{x}_a; \kappa^*) - \widetilde{m}_{\varepsilon,c}\|_{\widetilde{\mathcal{C}}_{\varepsilon,c}^{-1}}^2\right). \quad (4.14)$$

In order to obtain both a reconstruction as well as credibility intervals for the approximate parameter values, we approximate the *conditional mean* (CM) estimate

$$\mathbf{x}_{a,\mathrm{CM}} = \int_\mathcal{A} \mathbf{x}_a \pi(\mathbf{x}_a|\mathbf{y}) \, \mathrm{d}\mathbf{x}_a \quad (4.15)$$

via MCMC sampling of the second stage posterior distribution

$$\hat{\mathbf{x}}_{a,\mathrm{CM}} = \frac{1}{K}\sum_{k=1}^{K} \mathbf{x}_a^{(k)}, \quad (4.16)$$

where $\{\mathbf{x}_a^{(k)}\}_{k=1}^{\infty}$ is an ergodic Markov chain with invariant probability density $\pi(\mathbf{x}_a|\mathbf{y})$ obtained from the Metropolis-Hastings method which is described in Algorithm 6.

Algorithm 6 Random walk Metropolis-Hastings method

Require: initial guess $\mathbf{x}_a^{(1)}$, proposal covariance $\mathcal{C}_{\text{RW}}$, number of samples K

    for $k = 1 : K$ **do**

        Draw candidate $\tilde{\mathbf{x}}_a \leftarrow \mathbf{x}_a^{(k)} + \gamma_1$ with $\gamma_1 \sim \mathcal{N}(0, \mathcal{C}_{\text{RW}})$

        Calculate acceptance ratio $a(\mathbf{x}_a^{(k)}, \tilde{\mathbf{x}}_a) \leftarrow \min\left\{1, \frac{\pi(\tilde{\mathbf{x}}_a|\mathbf{y})}{\pi(\mathbf{x}_a^{(k)}|\mathbf{y})}\right\}$

        Draw $\gamma_2 \sim \mathcal{U}[0, 1]$

        if $a(\mathbf{x}_a^{(k)}, \tilde{\mathbf{x}}_a) \geq \gamma_2$ **then**

            Set $\mathbf{x}_a^{(k+1)} \leftarrow \tilde{\mathbf{x}}_a$

        else

            Set $\mathbf{x}_a^{(k+1)} \leftarrow \mathbf{x}_a^{(k)}$

        end if

    end for

    return $\mathbf{x}_a^{(1)}, ..., \mathbf{x}_a^{(K)}$

4.4 Numerical experiments

In this section, we provide numerical experiments to demonstrate the feasibility of our approach. In these experiments, the full computational domain was given by a polygonal approximation of D, still denoted D for convenience, with $R = 10$, while the truncated computational domain D_c used for the reconstruction method was rectangular. The anomaly Σ was assumed to be a circle of radius $r = 0.65$ centered at the point $(c_1, c_2) = (-1.825, -2.625)$, see Figure 4.1. Eight evenly spaced electrodes, each of length 0.65, were placed around the origin on $\partial_1 D$ and the contact impedance was assumed to be equal to 0.01 and known. Moreover, we assumed that the distribution of the underlying random field, as well as the scaling parameter ε and the distribution of the measurement noise are known. Finally, we also assumed that we know a priori we are searching for a circular, perfectly conducting anomaly.

In the reconstruction process we used three different triangulations, the finest being the triangulation $\mathcal{T}^h$ of D for the sampling of the homogenization error which consisted of 481 357 nodes and 947 676 triangles. In the first

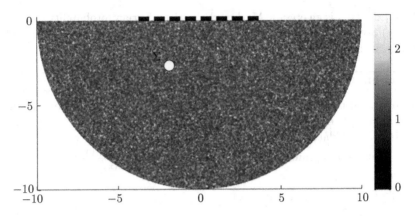

Figure 4.1: Basic geometric setting and measurement configuration with 8
electrodes; a realization of the random background conductivity is
shown and the anomaly is depicted in white.

stage, we used a simplistic rectangular grid $\mathcal{T}^{h_c}$ consisting of 600 nodes and
1104 triangles in order to keep the number of unknowns reasonably small.
In the second stage, where it is clearly important to interpolate Σ into the
triangulation with sufficient accuracy, we utilized a refined rectangular grid
$\mathcal{T}^{h'_c}$ with 1681 nodes and 3200 triangles.

The program code is implemented in MATLAB R2013A and uses meshing
and FEM routines from the EIDORS package, cf. [161], as well as the
HLIBpro library, cf. [22, 59]. The numerical experiments were performed on
a notebook with 2.3 GHz i7 quad core CPU equipped with 16 GB of RAM.

4.4.1 Synthetic measurement data

We considered a conductivity random field of the form

$$\kappa(x,\omega) = \exp(\tanh(G(x,\omega))), \qquad (4.17)$$

where $\{G(x,\omega), (x,\omega) \in \mathbb{R}^d \times \Gamma\}$ is a Gaussian random field with squared
exponential covariance function

$$\mathcal{C}_G^\infty(x,y) = \sigma_G^2 \exp\left(-\frac{|x-y|^2}{2\lambda^2}\right),$$

variance $\sigma_G^2 = 0.05$ and correlation length $\lambda = 0.003$. Obviously, the
conductivity random field satisfies assumption (A1) from Subsection 1.3.1.

Realizations of the underlying Gaussian random field were generated via circulant embedding of the covariance matrix, cf. [41]. See Appendix B for a description of this technique.

In order to avoid an inverse crime, the simulated noiseless voltage vector was computed via a finite element method using standard linear basis functions on the refined triangulation $\mathcal{T}^{h/2}$. Subsequently, the measurement vector $\mathbf{y}$ was generated by adding centered Gaussian noise with covariance matrix $\mathcal{C} = 5 \times 10^{-5} I_{N(N-1)}$, which corresponds to a mean noise level in the relative electrode voltage data of order

$$
\left\langle \frac{|\mathbf{E}|}{|\mathbf{U}(\kappa_\varepsilon|_{D\backslash\overline{\Sigma}}) - \mathbf{U}(\kappa_\varepsilon|_D)|} \right\rangle \approx 6\%,
$$

where $\langle \cdot \rangle$ denotes the sample average. The synthetic measurement data was generated by applying trigonometric current patterns $J^{(k)}$, $k = 1, ..., 7$, given by

$$
J_l^{(k)} = \begin{cases} \cos(k\theta_l), & k = 1, 2, ..., 4, \\ \sin((k-4)\theta_l), & k = 4, ..., 7, \end{cases}
$$

where $\theta_l = 2\pi l/8$, $l = 1, ..., 8$.

4.4.2 Precomputations

The key ingredient for the proposed reconstruction method is an accurate estimate $\hat{\kappa}_T^*$ of the effective conductivity κ^*. In order to obtain such an estimate, we used the continuum micro-scale Monte Carlo method described in the previous chapter. For the computation of κ^* we used 10^6 sample paths and the time horizon $T = 10$.

Subsequently, the mean m_ε and covariance $\mathcal{C}_\varepsilon$ of the enhanced error model were approximated via Monte Carlo sampling using the forward models $F_{h,D}$ and $\widetilde{F}_{h,D}$, respectively. Note that in each step of the Monte Carlo sampling scheme, a huge linear system has to be solved which is clearly the bottleneck of the computation. In order to cope with the computational complexity, we used the conjugate gradient method in conjunction with an LU preconditioner based on low-rank approximation by hierarchical matrices. Indeed it has been shown that $\mathcal{H}$-LU factorization yields problem-independent convergence rates and may thus be used as a black-box preconditioner for iterative methods, cf. [17, 59]. In our case, the sparse stiffness matrix of the discretized problem was written into the appropriate data structure of the HLIBpro library and transformed into an $\mathcal{H}$-matrix. No geometrical data was assumed and hence,

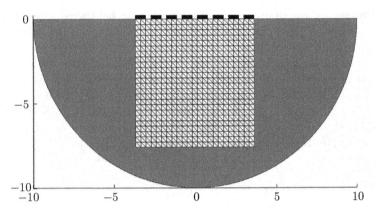

Figure 4.2: Computational domains D and D_c with the corresponding triangulations $\mathcal{T}^h$ and $\mathcal{T}^{h_c}$, respectively. Note that the individual triangles of the fine triangulation $\mathcal{T}^h$ can not be depicted.

algebraic clustering was performed for the conversion. Furthermore, nested dissection was applied to improve the efficiency of the $\mathcal{H}$-LU factorization.

For the computation of the approximation error statistics for the reduced forward models, F_{h_c, D_c} and $\widetilde{F}_{h'_c, D_c}$, we proceeded accordingly. Both reduced forward maps were computed on the truncated rectangular domain D_c depicted in Figure 4.2 with boundary condition

$$\hat{\kappa}_T^* \nu \cdot \nabla u|_{\partial D} + g u|_{\partial D} = f \quad \text{on } \partial D_c.$$

Each enhanced error model was obtained from Monte Carlo estimates based on 10 000 independent draws of the random field, respectively the anomaly prior.

4.4.3 Results and discussion

For the given random field (4.17), we computed the approximate effective conductivity $\hat{\kappa}_T^* = 1.026 \pm 9 \times 10^{-3}$ in the precomputation step (we rounded to 3 digits).

In the first stage, we computed three iterations of the Gauss-Newton algorithm, where the homogenized conductivity $\hat{\kappa}_T^*$ was used as initial guess. The matrix A, the gradient and the Jacobian of the function $\Phi_\alpha(\mathbf{x}_c)$ were computed using the standard finite element discretization on $\mathcal{T}^{h_c}$ and the regularization parameter α was chosen ad-hoc. Moreover, the distribution of the measurement noise process as well as the contact impedance were assumed to be known a priori.

In the second stage, a conditional mean estimate for the bootstrap prior was computed using the random walk Metropolis-Hastings algorithm. The proposal covariance $\mathcal{C}_{RW}$ was chosen after a number of calibration runs and comparison of the integrated autocorrelation times of the corresponding Markov chains, cf. [113]. We used the refined finite element discretization on $\mathcal{T}^{h'_c}$ in this stage. The sampling run was of length 100 000 steps, where the first 2000 steps were discarded as burn-in steps.

In order to demonstrate the advantage of the enhanced error model accounting for the homogenization error, we considered two experiments which differed only with regard to the latter. In Experiment 1, the error caused by using a homogenized forward model was ignored, while the approximation errors caused by the coarse finite element discretization and truncation of the domain were accounted for as described in Subsection 4.2.2. In Experiment 2, the enhanced error model accounting for the homogenization error introduced in Subsection 4.2 was included into the approximation error model.

It turned out that the mean of the homogenization error in the second experiment was significantly larger than the measurement error, indicating that the test setting at hand can not at all be considered a *homogenization regimen*, which would render additional error modeling redundant. To be precise, we observed that

$$\left\langle \frac{|\mathbf{U}(\kappa_\varepsilon|_{D\backslash\overline{\Sigma}}) - \overline{\mathbf{U}}(\hat{\kappa}_T^*|_{D\backslash\overline{\Sigma}})|}{|\mathbf{U}(\kappa_\varepsilon|_{D\backslash\overline{\Sigma}}) - \mathbf{U}(\kappa_\varepsilon|_D)|} \right\rangle \approx 33\%.$$

For the first experiment, the MAP estimates obtained in the first stage for parameter values $\alpha = 10^{-4}$ and $\alpha = 10^{-2}$ are shown in Figure 4.3 on the next page and for the second experiment, the MAP estimates obtained in the first stage for parameter values $\alpha = 10^{-4}$ and $\alpha = 10^{-2}$ are shown in Figure 4.4; the location of the anomaly is also shown for comparison. Obviously, ignoring the homogenization error has lead to severe reconstruction artefacts, in particular it is not possible to locate the anomaly from the MAP estimate in a reliable manner. On the other hand, it can be seen that using the enhanced error model accounting for the homogenization error improved the quality of the MAP estimate significantly. Moreover, it can be observed that decreasing the parameter led to better localization of the anomaly.

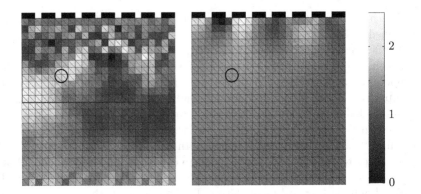

Figure 4.3: Experiment 1 (conventional error model): MAP estimates $\mathbf{x}_{c,\mathrm{MAP}}$ corresponding to parameters $\alpha = 10^{-4}$ (left) and $\alpha = 10^{-2}$ (right). The ROI is shown as a rectangle; the position of the anomaly is indicated by the circle.

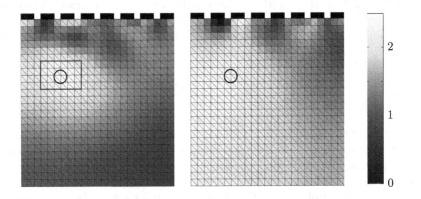

Figure 4.4: Experiment 2 (enhanced error model): MAP estimates $\mathbf{x}_{c,\mathrm{MAP}}$ corresponding to parameters $\alpha = 10^{-4}$ (left) and $\alpha = 10^{-2}$ (right). The ROI is shown as a rectangle; the position of the anomaly is indicated by the circle.

Although we were able to reduce electrode artefacts by using a smoothness prior with Dirichlet boundary condition, reconstruction errors still occurred near the electrodes. Nevertheless, in summary it can be stated that model reduction errors due to homogenization, coarse finite element discretization and truncation of the computational domain can be reduced simultaneously via approximation of the first two moments of the resulting errors. Indeed, the estimate corresponding to $\alpha = 10^{-4}$ gives quite precise information about the center of the anomaly but not about its size. Moreover, we could confirm the observation made in [6, 100] concerning the robustness of the approximation error approach with respect to misspecified priors. The computation time for the MAP estimates was around 120 seconds.

In both experiments, the threshold parameter was chosen by visual inspection and the corresponding ROIs for the parameters c_1 and c_2 are plotted in the left-hand side of Figure 4.3 and Figure 4.4, respectively. The parameter bounds for these parameters used in the second stage bootstrap prior corresponded to the depicted ROIs. For the parameter r, the bounds 0.45 and 0.9 were chosen in both experiments. The concrete values of the parameter bounds correspond to the axis limits in Figure 4.5 and Figure 4.7, respectively.

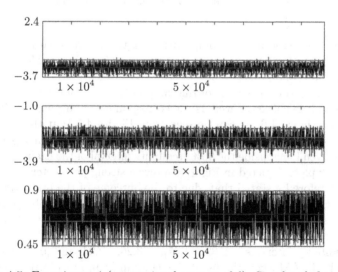

Figure 4.5: Experiment 1 (conventional error model): Simulated chains for the parameters c_1, c_2 and r after the burn-in phase. The solid line shows the CM estimator, the dashed lines show the 90% credibility interval. The light solid line shows the exact solution

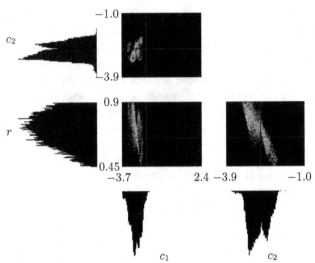

Figure 4.6: Experiment 1 (conventional error model): Joint posterior densities
and marginal densities obtained from a sampling run of length
100 000 steps, where the first 2000 steps were discarded as burn-in
steps. Crosshairs point to the true parameter values.

In the first experiment, computation of the approximate conditional
mean estimate (4.16) gave $\hat{\mathbf{x}}_{a,\mathrm{CM}} = (-2.67, -2.72, 0.71)^T$. Moreover, the
90% credibility intervals $[-3.32, -2.07]$, $[-3.19, -2.17]$ and $[0.51, 0.87]$ were
obtained. The Markov chains of the sampling run are plotted in Figure 4.5
and the joint densities as well as the corresponding marginal densities are
plotted in Figure 4.6. It can be seen from Figure 4.5 that the horizontal
position c_1 of the anomaly is greatly misspecified with the exact parameter
value not even lying inside the 90% credibility interval of the chain. Moreover,
the density plots depicted in Figure 4.6 reveal strong parameter correlations.
It can therefore be stated that, due to the outcome of the first experiment,
ignoring the homogenization error is not advisable in the problem at hand.

In the second experiment, computation of the approximate conditional
mean estimate (4.16) gave $\hat{\mathbf{x}}_{a,\mathrm{CM}} = (-1.87, -2.74, 0.69)^T$. Moreover, the
90% credibility intervals $[-2.32, -1.47]$, $[-3.26, -2.24]$ and $[0.54, 0.84]$ were
obtained. The Markov chains of the sampling run are plotted in Figure
4.7 and the joint densities as well as the corresponding marginal densities
are plotted in Figure 4.8. With regard to the fact that measurements were
only taken on the accessible part of the boundary, the results obtained

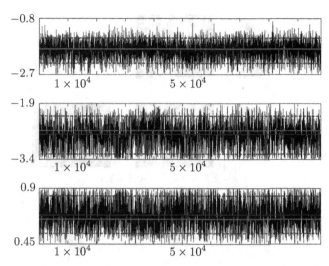

Figure 4.7: Experiment 2 (enhanced error model): Simulated chains for the parameters c_1, c_2 and r after the burn-in phase. The solid line shows the CM estimator, the dashed lines show the 90% credibility interval. The light solid line shows the exact solution.

using the enhanced error model accounting for the homogenization error are encouraging. Both, the size and the position of the anomaly were located with reasonable accuracy and all three chains seem to be close to their stationary distribution. This is partly due to the fact, that the anomaly was a priori known to be a perfect conductor. Indeed, correlations between the unknown parameters in the vector $\mathbf{x}_a$ were far less severe than the correlations observed in [139] between these parameters and an unknown conductivity. At the same time this observation yields an explanation of the good performance of the standard random walk Metropolis-Hastings algorithm, which is known to perform rather poor when parameters are strongly correlated. Moreover, a comparison of Figure 4.6 and Figure 4.8 indicates that the restriction to the narrow box obtained from the ROI in the second experiment could prevent the random walk from getting stuck around local maxima of the posterior density. The computation time for 10 000 Monte Carlo samples from the second stage posterior distribution was around 48 minutes.

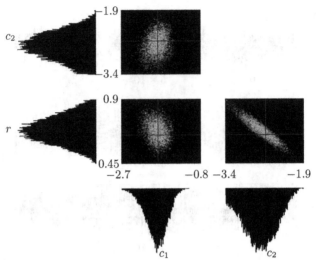

Figure 4.8: Experiment 2 (enhanced error model): Joint posterior densities
and marginal densities obtained from a sampling run of length
100 000 steps, where the first 2000 steps were discarded as burn-in
steps. Crosshairs point to the true parameter values.

Conclusion

To sum up, it can be stated that the proposed method provides reasonable
results for the test case at hand. We found that the quality of the reconstruction is similar to that of the reconstructions in [139], where a similar
two-stage method was applied to the deterministic problem of locating an
anomaly (with known finite conductivity) in a homogeneous background
conductivity. Future work will concentrate on numerical methods to speed
up the time-consuming precomputation step as well as on problems, where
less prior information about the random background medium is available.
For instance, problems where the scaling parameter ε is unknown and must
therefore also be inferred from the measured data.

Appendices

A Basic Dirichlet form theory

In this chapter, we collect some results from the theory of symmetric Dirichlet forms for further reference. Our primary sources are the monographs [54] by Fukushima, Ōshima and Takeda and [115] by Ma and Röckner.

Symmetric forms, semigroups and resolvents

Let E be a locally compact separable metric space and let the *reference measure* m be a positive Radon measure on E with full support. We consider the σ-finite measure space $(E, \mathcal{B}(E), m)$, where $\mathcal{B}(E)$ denotes the Borel σ-algebra on E. Let $L^2(E; m)$ denote the standard Lebesgue space consisting of square integrable, m-measurable real-valued functions on E. The inner product and norm on $L^2(E; m)$ are denoted $\langle \cdot, \cdot \rangle$ and $\|\cdot\|$, respectively. For a function $v \in L^2(E; m)$, $\mathrm{supp}(v)$ is defined to be the support of the measure $v \cdot m$. As usual, we set for $v, w \in L^2(E; m)$

$$v \vee w := \sup\{v, w\}, \quad v \wedge w := \inf\{v, w\}$$

and write $v \leq w$ or $v < w$ if the inequalities hold m-a.e. for corresponding representatives.

Definition A.1. The pair $(\mathcal{E}, \mathcal{D}(\mathcal{E}))$ is a *symmetric form* on $L^2(E; m)$ if $\mathcal{D}(\mathcal{E})$ is a dense linear subspace of $L^2(E; m)$ and $\mathcal{E}$ is a non-negative definite symmetric bilinear form on $\mathcal{D}(\mathcal{E}) \times \mathcal{D}(\mathcal{E})$.

For each $\alpha > 0$ we define

$$\mathcal{E}_\alpha(v, w) := \mathcal{E}(v, w) + \alpha \langle v, w \rangle, \quad v, w \in \mathcal{D}(\mathcal{E}),$$

and call $(\mathcal{E}, \mathcal{D}(\mathcal{E}))$ *closed* if $\mathcal{D}(\mathcal{E})$ is complete with respect to the norm

$$\|\cdot\|_{\mathcal{E}_\alpha} := \sqrt{\mathcal{E}_\alpha(\cdot, \cdot)}.$$

Definition A.2. A closed symmetric form $(\mathcal{E}, \mathcal{D}(\mathcal{E}))$ on $L^2(E; m)$ is said to be a *symmetric Dirichlet form* if the *unit contraction* operates on $(\mathcal{E}, \mathcal{D}(\mathcal{E}))$, i.e., given $v \in \mathcal{D}(\mathcal{E})$,

$$w := (0 \vee v) \wedge 1 \in \mathcal{D}(\mathcal{E}) \text{ and } \mathcal{E}(w, w) \leq \mathcal{E}(v, v).$$

Definition A.3. A family of symmetric linear operators $\{T_t, t \geq 0\}$ on $L^2(E; m)$ is called a *strongly continuous contraction semigroup* if for every $v \in L^2(E; m)$ and for all $s, t \geq 0$ the following properties hold:

(i) $T_s T_t v = T_{s+t} v$;

(ii) $||T_t v|| \leq ||v||$;

(iii) $\lim_{t \to 0+} ||T_t v - v|| = 0$.

Definition A.4. A family of symmetric linear operators $\{G_\alpha, \alpha > 0\}$ on $L^2(E; m)$ is called a *strongly continuous contraction resolvent* if for every $v \in L^2(E; m)$ and for all $\alpha, \beta > 0$ the following properties hold:

(i) $G_\alpha v - G_\beta v + (\alpha - \beta) G_\alpha G_\beta v = 0$;

(ii) $\alpha ||G_\alpha v|| \leq ||v||$;

(iii) $\lim_{\alpha \to \infty} ||\alpha G_\alpha v - v|| = 0$.

The *infinitesimal generator* of a strongly continuous contraction resolvent $\{G_\alpha, \alpha > 0\}$ on $L^2(E; m)$ is defined by

$$\mathcal{L}v = \alpha v - G_\alpha^{-1} v, \quad \mathcal{D}(\mathcal{L}) = G_\alpha(L^2(E; m))$$

and $\mathcal{L}$ is a non-positive definite self-adjoint operator, independent of α. The *infinitesimal generator* of a strongly continuous contraction semigroup $\{T_t, t \geq 0\}$ on $L^2(E; m)$ is defined by

$$\mathcal{L}v = \lim_{t \to 0+} \frac{T_t v - v}{t}, \quad \mathcal{D}(\mathcal{L}) = \{v \in L^2(E; m) : \mathcal{L}v \text{ exists}\}$$

and $(\mathcal{L}, \mathcal{D}(\mathcal{L}))$ is easily seen to coincide with the generator of the resolvent of $\{T_t, t \geq 0\}$ given by the Laplace transform

$$G_\alpha v = \int_0^\infty e^{-\alpha t} T_t v \, dt, \quad \alpha > 0.$$

Let $\{E_\lambda, \lambda \in \mathbb{R}\}$ denote a spectral family and ϕ a continuous real-valued function on $\mathbb{R}$. Then there exists a unique self-adjoint operator $\mathcal{L} = \int_{-\infty}^\infty \phi(\lambda) \, dE_\lambda$ on $L^2(E; m)$ such that

$$\langle \mathcal{L}v, w \rangle = \int_{-\infty}^\infty \phi(\lambda) \, d(E_\lambda v, w)$$

and

$$\mathcal{D}(\mathcal{L}) = \left\{ v \in L^2(E; m) : \int_{-\infty}^{\infty} \phi(\lambda)^2 \, d(E_\lambda v, v) < \infty \right\}.$$

For every non-positive definite self-adjoint operator $\mathcal{L}$, the operator $-\mathcal{L}$ admits a unique spectral family $\{E_\lambda, \lambda \in \mathbb{R}\}$ with $E_\lambda = 0$ for every $\lambda < 0$, cf. [162]. In particular, the square root of $-\mathcal{L}$ is well-defined by

$$\sqrt{-\mathcal{L}} = \int_0^{\infty} \sqrt{\lambda} \, dE_\lambda.$$

Definition A.5. A family of linear operators $\{R_t, t \geq 0\}$ on $L^2(E; m)$ is called *sub-Markovian* if the property $0 \leq v \leq 1$ m-a.e. implies that $0 \leq R_t v \leq 1$ m-a.e. for all $t \geq 0$.

A proof of the following fundamental theorem can be found in [54, Chapter 1]. We confine ourselves here to collecting the underlying relations in Figure 1.1 below.

Theorem A.6. *The following objects are mutually in one-to-one correspondence:*

(i) *The set of symmetric Dirichlet forms $(\mathcal{E}, \mathcal{D}(\mathcal{E}))$ on $L^2(E; m)$;*

(ii) *the set of non-positive definite self-adjoint operators $(\mathcal{L}, \mathcal{D}(\mathcal{L}))$ on $L^2(E; m)$ such that the family $\{\exp(t\mathcal{L}), t \geq 0\}$ is sub-Markovian;*

(iii) *the set of strongly continuous sub-Markovian contraction semigroups $\{T_t, t \geq 0\}$ on $L^2(E; m)$;*

(iv) *the set of strongly continuous contraction resolvents $\{G_\alpha, \alpha > 0\}$ on $L^2(E; m)$ such that the family $\{\alpha G_\alpha, \alpha > 0\}$ is sub-Markovian.*

Definition A.7. A Dirichlet form $(\mathcal{E}, \mathcal{D}(\mathcal{E}))$ on $L^2(E; m)$ is said to be *regular* if $\mathcal{D}(\mathcal{E}) \cap C_c(E)$ is dense in both, $(\mathcal{D}(\mathcal{E}), \|\cdot\|_{\mathcal{E}_1})$ and $(C_c(E), \|\cdot\|_\infty)$.

Definition A.8. A symmetric Dirichlet form $(\mathcal{E}, \mathcal{D}(\mathcal{E}))$ on $L^2(E; m)$ is called *local* if $\mathcal{E}(v, w) = 0$ whenever $v, w \in \mathcal{D}(\mathcal{E})$ have disjoint compact supports and it is called *strongly local* if the same holds true, whenever w is constant on a neighborhood of $\operatorname{supp}(v)$.

Let $\{T_t, t \geq 0\}$ denote a sub-Markovian strongly continuous contraction semigroup on $L^2(E; m)$. Then we define for $v, w \in L^2(E; m)$ the *approximating symmetric form*

$$\mathcal{E}^{(t)}(v, w) := t^{-1} \langle v - T_t v, w \rangle, \quad t > 0,$$

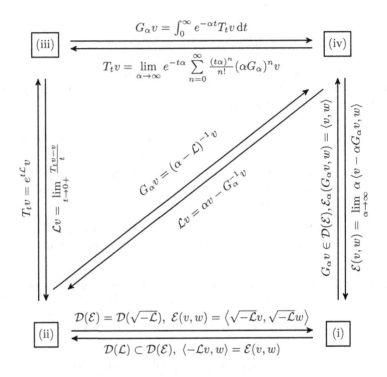

Figure 1.1: The one-to-one correspondences from Theorem A.6

and the following lemma provides us with a simple direct description of the associated Dirichlet form in terms of the semigroup.

Lemma A.9. *For $v \in L^2(E; m)$, $\mathcal{E}^{(t)}(v, v)$ is non-decreasing as $t \to 0+$ and the symmetric Dirichlet form associated with $\{T_t, t \geq 0\}$ is given by*

$$\mathcal{E}(v, w) = \lim_{t \to 0+} \mathcal{E}^{(t)}(v, w), \quad \mathcal{D}(\mathcal{E}) = \{v \in L^2(E; m) : \lim_{t \to 0+} \mathcal{E}^{(t)}(v, v) < \infty\}.$$

Potential theory for regular Dirichlet forms

From now on let $(\mathcal{E}, \mathcal{D}(\mathcal{E}))$ be a regular symmetric Dirichlet form on $L^2(E; m)$. It turns out that the Lebesgue measure is not fine enough to develop the potential theory for regular Dirichlet forms. Neither is it fine enough to characterize the stochastic processes which can be associated with regular symmetric Dirichlet forms. Appropriate $\mathcal{E}$-quasi-notions may be defined for Dirichlet forms on general topological spaces; in our setting these notions can be interpreted in terms of *capacities*. Denote $\mathcal{O}$ the set of all open subsets of E and define for $O \in \mathcal{O}$ the set $\mathcal{M}_O := \{v \in \mathcal{D}(\mathcal{E}) : v \geq 1 \ m\text{-a.e. on } O\}$.

Definition A.10. The *capacity* of an open set $O \in \mathcal{O}$ is defined as

$$\mathrm{Cap}(O) = \inf\{\mathcal{E}_1(v,v) : v \in \mathcal{M}_O\}$$

with the convention $\mathrm{Cap}(\emptyset) = \infty$ and for an arbitrary subset $B \subset E$

$$\mathrm{Cap}(B) = \inf\{\mathrm{Cap}(O) : O \in \mathcal{O} \text{ and } B \subset O\}.$$

$\mathrm{Cap}(\cdot)$ can be shown to be a Choquet capacity, i.e.,

$$\mathrm{Cap}(\cup_{k\in\mathbb{N}} B_k) = \sup_{k\in\mathbb{N}} \mathrm{Cap}(B_k), \quad \mathrm{Cap}(\cap_{k\in\mathbb{N}} K_k) = \inf_{k\in\mathbb{N}} \mathrm{Cap}(K_k)$$

for any increasing sequence $\{B_k\}_{k\in\mathbb{N}}$ of subsets of E and any decreasing sequence $\{K_k\}_{k\in\mathbb{N}}$ of compact subsets, cf. [54]. A Borel set $\mathcal{N}$ is called $\mathcal{E}$-*exceptional* if $\mathrm{Cap}(\mathcal{N}) = 0$. A statement depending on $x \in E$ is said to be true $\mathcal{E}$-*quasi-everywhere* (abbreviated "$\mathcal{E}$-q.e.") on E if there is an $\mathcal{E}$-exceptional set $\mathcal{N}$ such that the statement is true for every $x \in E \backslash \mathcal{N}$. Two functions are said to be $\mathcal{E}$-*quasi-equivalent* if they differ only on an exceptional set. A function is called $\mathcal{E}$-*quasi-continuous* if for every $\varepsilon > 0$, there exists an open set $O \in \mathcal{O}$ with $\mathrm{Cap}(O) < \varepsilon$ such that $v|_{E\backslash O}$ is finite and continuous. By [54, Theorem 2.1.3], every function in $\mathcal{D}(\mathcal{E})$ admits a quasi-continuous version if $(\mathcal{E}, \mathcal{D}(\mathcal{E}))$ is a regular Dirichlet form. Throughout this work, functions from the extended Dirichlet space of a regular Dirichlet form will be represented by their $\mathcal{E}$-quasi-continuous versions. Moreover, since in this work we are dealing with a fixed regular Dirichlet form, we usually drop the "$\mathcal{E}$-" from the notation.

Definition A.11. A positive Borel measure μ on $(E, \mathcal{B}(E))$ is called *smooth* (with respect to $(\mathcal{E}, \mathcal{D}(\mathcal{E}))$) if it satisfies the following conditions:

(i) μ charges no sets of zero capacity, i.e., $\mu(\mathcal{N}) = 0$ if $\mathcal{N}$ is exceptional;

(ii) there exists a nest $\{C_k\}_{k\in\mathbb{N}}$ such that $\mu(C_k) < \infty$ for every $k \in \mathbb{N}$.

We denote by $\mathcal{S}$ the family of all smooth positive Borel measures on $(E, \mathcal{B}(E))$.

Definition A.12. A positive Radon measure μ on $(E, \mathcal{B}(E))$ is said to have *finite energy* (with respect to $(\mathcal{E}, \mathcal{D}(\mathcal{E}))$) if there is a constant $c > 0$ such that

$$\int_E |v(x)| \, d\mu(x) \leq c \|v\|_{\mathcal{E}_1}, \quad v \in \mathcal{D}(\mathcal{E}) \cap C_c(E).$$

We denote by $\mathcal{S}_0$ the family of all positive Radon measures on $(E, \mathcal{B}(E))$ having finite energy.

Note that every finite Radon measure charging no sets of zero capacity is smooth, in particular $\mathcal{S}_0 \subset \mathcal{S}$.

For a regular symmetric Dirichlet form $(\mathcal{E}, \mathcal{D}(\mathcal{E}))$, every measure $\mu \in \mathcal{S}_0$ defines a bounded linear functional on the Hilbert space $(\mathcal{D}(\mathcal{E}), \mathcal{E}_\alpha)$, $\alpha > 0$, namely

$$v \mapsto \int_E v(x) \, d\mu(x).$$

By the Riesz representation theorem, μ has finite energy if and only if for each $\alpha > 0$ there exists a unique so-called α-*potential* $U_\alpha \mu \in \mathcal{D}(\mathcal{E})$ such that

$$\mathcal{E}_\alpha(U_\alpha \mu, v) = \int_E v(x) \, d\mu(x) \quad \text{for every } v \in \mathcal{D}(\mathcal{E}) \cap C_c(E).$$

Definition A.13. A function $v \in L^2(E; m)$ is called α-*excessive*, $\alpha > 0$, (with respect to $(\mathcal{E}, \mathcal{D}(\mathcal{E}))$) if $e^{-\alpha t} T_t v \leq v$ m-a.e. for every $t > 0$, where $\{T_t, t \geq 0\}$ denotes the strongly continuous contraction semigroup on $L^2(E; m)$ associated with $(\mathcal{E}, \mathcal{D}(\mathcal{E}))$.

According to the analytic theory of Beurling and Deny, cf. [20], every regular symmetric Dirichlet form on $L^2(E; m)$ admits a unique decomposition into a *diffusion part*, a *jumping part* and a *killing part*.

Theorem A.14. *Let $(\mathcal{E}, \mathcal{D}(\mathcal{E}))$ be a regular symmetric Dirichlet form on $L^2(E; m)$. Then $\mathcal{E}$ admits a unique decomposition*

$$\mathcal{E}(v, w) = \mathcal{E}^{(c)}(v, w) + \int_{E \times E \setminus \delta} (v(x) - v(y))(w(x) - w(y)) \, d\mathcal{J}(x, y)$$

$$+ \int_E vw \, d\mathcal{K}(x), \quad v, w \in \mathcal{D}(\mathcal{E}) \cap C_c(E).$$

The diffusion part $\mathcal{E}^{(c)}$ is a strongly local closed symmetric form. The jumping measure $\mathcal{J}$ is a symmetric Radon measure on $E \times E \setminus \delta$, δ being the diagonal set of $E \times E$, which satisfies $\mathcal{J}(E \times \mathcal{N}) = 0$ if $\mathrm{Cap}(\mathcal{N}) = 0$. The killing measure $\mathcal{K}$ is a positive Radon measure from $\mathcal{S}_0$.

Regular Dirichlet forms and symmetric Hunt processes

Definition A.15. A function $P : E \times \mathcal{B}(E) \to [0,1]$ is called a *probability transition kernel* on $(E, \mathcal{B}(E))$ if

(i) for each $x \in E$, the mapping $B \mapsto P(x, B)$ is a probability measure on $(E, \mathcal{B}(E))$;

(ii) for each $B \in \mathcal{B}(E)$, the mapping $x \mapsto P(x, B)$ is a measurable function.

If the mapping in (i) is not a probability measure but a positive measure satisfying $P(x, E) \leq 1$ for each $x \in E$, then P is called a *sub-Markovian transition kernel*; if it is merely a positive measure, then P is simply called a *kernel*.

Remark A.16. Note that a sub-Markovian transition kernel P may always be extended to a probability transition kernel P_{∂} by switching to the one-point compactification $E_{\partial} := E \cup \{\partial\}$ and assigning the missing mass to the *cemetery point* ∂ via $P_{\partial}(x, \{\partial\}) = 1 - P(x, E)$ and $P_{\partial}(\partial, \{\partial\}) = 1$. If E is already compact, then ∂ is adjoined as an isolated point. The underlying σ-field is extended to $\sigma(\mathcal{B}(E), \{\partial\})$. We adopt the convention to extend both, real-valued functions defined on E as well as the reference measure m to E_{∂} by zero. Throughout this work we will always assume that the state space possesses a cemetery point, although we will usually omit the subscript "∂" in our notation.

A transition kernel acts from the left on the set of bounded Borel functions by

$$P\phi(x) = \int_E \phi(y) P(x, \mathrm{d}y)$$

and it also acts from the right on the set of probability measures on $(E, \mathcal{B}(E))$ by

$$\mu P(B) = \int_E P(x, B) \mu(\mathrm{d}x).$$

Let $X = (\Omega, \mathcal{F}, \{X_t, t \geq 0\}, \mathbb{P}_x)$ be a stochastic process with state space $(E, \mathcal{B}(E))$, i.e., $(\Omega, \mathcal{F}, \mathbb{P}_x)$ is a probability space and $X_t : \Omega \to E$ is an $(\mathcal{F}\text{-}\mathcal{B}(E))$-measurable map. We say that a family $\{\tilde{\mathcal{F}}_t, t \geq 0\}$ of sub-σ-algebras of $\mathcal{F}$ is an *admissible filtration* if $\tilde{\mathcal{F}}_t$ is increasing with respect to t and for all $t \geq 0$, X_t is a measurable map with respect to the σ-algebras $\tilde{\mathcal{F}}_t$ and $\mathcal{B}(E)$. An admissible filtration is *right-continuous* if $\tilde{\mathcal{F}}_t = \cap_{s>t} \mathcal{F}_s$ for all $t \geq 0$. We set $\mathcal{F}_\infty := \sigma(X_s, 0 \leq s < \infty)$ and $\mathcal{F}_t := \sigma(X_s, s \leq t)$, $0 \leq t < \infty$, and we call the family $\{\mathcal{F}_t, t \geq 0\}$ the *minimum admissible filtration*.

Definition A.17. A quadruplet $X = (\Omega, \mathcal{F}, \{X_t, t \geq 0\}, \mathbb{P}_x)$ is called a *Markov process* with state space $(E, \mathcal{B}(E))$ if the following conditions are satisfied:

(i) For each $x \in E$, the quadruplet $(\Omega, \mathcal{F}, \{X_t, t \geq 0\}, \mathbb{P}_x)$ is a stochastic process with state space $(E, \mathcal{B}(E))$ and $X_\infty(\omega) = \partial$ for all $\omega \in \Omega$;

(ii) for each $t \geq 0$ the function $E \times \mathcal{B}(E) \to [0, 1]$, $(x, B) \mapsto \mathbb{P}_x(X_t \in B)$ is a probability transition kernel;

(iii) there exists an admissible filtration $\{\mathcal{F}_t, t \geq 0\}$ such that for all $x \in E$, $B \in \mathcal{B}(E)$ and for all $s, t \geq 0$

$$\mathbb{P}_x\{X_{s+t} \in B | \mathcal{F}_t\} = \mathbb{P}_{X_t}\{X_s \in B\}, \quad \mathbb{P}_x\text{-a.s.};$$

(iv) $\mathbb{P}_x\{X_0 = x\} = 1$ for all $x \in E$;

(v) $\mathbb{P}_\partial\{X_t = \partial\} = 1$ for all $t \geq 0$.

The *lifetime* of the Markov process X is given by the random variable

$$\zeta(\omega) = \inf\{t \in [0, \infty] : X_t(\omega) = \partial\}.$$

If $\mathbb{P}_x\{\zeta = \infty\} = 1$ for all $x \in E$, then X is called *conservative*.

Definition A.18. A family $\{P_t, t \geq 0\}$ is called a *transition semigroup* on $(E, \mathcal{B}(E))$ if for each $t \geq 0$, P_t is a probability transition kernel on $(E, \mathcal{B}(E))$ satisfying the following conditions:

(i) $P_s P_t \phi = P_{s+t} \phi$ for $s, t \geq 0$ and every bounded Borel function ϕ;

(ii) for each $B \in \mathcal{B}(E)$, the mapping $(t, x) \mapsto P_t(x, B)$ is measurable with respect to $\mathcal{B}([0, \infty)) \times \mathcal{B}(E)$;

(iii) for each $x \in E$, $P_0(x, \cdot) = \delta_x(\cdot)$, where δ_x denotes the unit mass concentrated in x;

(iv) $\lim_{t \to 0+} P_t v(x) = v(x)$ for every bounded continuous function v and all $x \in E$.

By the *transition semigroup of a Markov process* X we mean the transition semigroup $\{P_t, t \geq 0\}$ which is determined by

$$P_t\phi(x) = \mathbb{E}_x\phi(X_t), \quad t \geq 0, \ x \in E$$

for every non-negative Borel function ϕ.

Definition A.19. A *Feller process* is a Markov process whose transition semigroup $\{P_t, t \geq 0\}$ is sub-Markovian, has the *Feller property*, i.e., $P_t\phi \in C^\infty(E)$ for all $\phi \in C^\infty(E)$ and all $t \geq 0$, and which is strongly continuous on $C^\infty(E)$, i.e., $\lim_{t \to 0+} \|P_t\phi - \phi\|_\infty = 0$ for all $f \in C^\infty(E)$.

Definition A.20. A sub-Markovian semigroup $\{T_t, t \geq 0\}$ on $L^2(E; m)$ is called *m-symmetric* if for every $t \geq 0$ and all non-negative Borel functions ϕ, ψ

$$\int_E (T_t\phi)(x)\psi(x) \, dm(x) = \int_E \phi(x)(T_t\psi)(x) \, dm(x).$$

Definition A.21. A sub-Markovian semigroup $\{T_t, t \geq 0\}$ on $L^2(E; m)$ is said to have the *strong Feller property* if $T_t\phi \in C_b(E)$ for all bounded Borel functions ϕ and all $t \geq 0$.

Remark A.22. Note that the strong Feller property does not necessarily entail the Feller property.

Unlike Feller processes, *Hunt processes* must be defined on the level of sample paths.

Definition A.23. A Markov process X with state space $(E, \mathcal{B}(E))$ is called a *Hunt process* if it satisfies the following conditions:

(i) For each $\omega \in \Omega$, the sample path $t \mapsto X_t(\omega)$ is right-continuous on $[0, \infty)$ and has left limits on $(0, \infty)$ in E;

(ii) there exists a right-continuous admissible filtration $\{\mathcal{F}_t, t \geq 0\}$ such that for any $\{\mathcal{F}_t\}$-stopping time τ and for every probability measure μ on $(E, \mathcal{B}(E))$ and $B \in \mathcal{B}(E)$

$$\mathbb{P}_\mu\{X_{\tau+s} \in B | \mathcal{F}_\tau\} = \mathbb{P}_{X_\tau}\{X_s \in B\}, \quad \mathbb{P}_\mu\text{-a.s. for all } s \geq 0;$$

(iii) for each $\omega \in \Omega$, the sample path $t \mapsto X_t(\omega)$ is *quasi-left-continuous* on $(0, \infty)$, i.e., for any family $\{\tau_k\}_{k \in \mathbb{N}}$ of $\{\mathcal{F}_t\}$-stopping times increasing to τ and for every probability measure μ on $(E, \mathcal{B}(E))$ we have

$$\mathbb{P}_\mu\left\{ \lim_{k \to \infty} X_{\tau_k} = X_\tau, \tau < \infty \right\} = \mathbb{P}_\mu\{\tau < \infty\}.$$

The following theorem was proved by Fukushima in 1971 in the seminal paper [52].

Theorem A.24. *Let* $(\mathcal{E}, \mathcal{D}(\mathcal{E}))$ *be a regular symmetric Dirichet form on* $L^2(E; m)$. *Then there exists a Hunt process* X *with state space* $(E, \mathcal{B}(E))$ *possessing an* m-*symmetric transition semigroup* $\{P_t, t \geq 0\}$ *so that the strongly continuous contraction semigroup* $\{T_t, t \geq 0\}$ *on* $L^2(E; m)$ *associated with* $(\mathcal{E}, \mathcal{D}(\mathcal{E}))$ *is related to* $\{P_t, t \geq 0\}$ *by the equality*

$$T_t \phi = P_t \phi \quad m\text{-}a.e. \text{ for all } t \geq 0$$

and for every non-negative Borel function ϕ.

In the situation of Theorem A.24 we say that the operator $(\mathcal{L}, \mathcal{D}(\mathcal{L}))$ associated with $(\mathcal{E}, \mathcal{D}(\mathcal{E}))$ is the *infinitesimal generator* of the m-symmetric Hunt process X. This Hunt process is unique up to the following equivalence.

Theorem A.25. *Let* $X^{(1)}$ *and* $X^{(2)}$ *be* m-*symmetric Hunt processes with state space* $(E, \mathcal{B}(E))$ *associated with the same regular symmetric Dirichlet form* $(\mathcal{E}, \mathcal{D}(\mathcal{E}))$ *on* $L^2(E; m)$. *Then there exists an exceptional set* $\mathcal{N}$, *common for both processes, such that their transition semigroups coincide on* $E \backslash \mathcal{N}$. *In this case,* $X^{(1)}$ *and* $X^{(2)}$ *are called* equivalent.

Next, let us recall the definition of *additive functionals* of a Hunt process X depending on the potential theory of the underlying Dirichlet form. The *time shift operator* Θ on the sample space Ω is defined by $X_s(\Theta_t \omega) = X_{t+s}(\omega)$, $s, t \geq 0$.

A real-valued stochastic process $A = (\Omega, \mathcal{F}, \{A_t, t \geq 0\}, \mathbb{P}_x)$ is an *additive functional* of X if the following conditions hold:

(i) there exists a *defining set* $\Lambda \in \mathcal{F}_\infty$ and an $\mathcal{E}$-exceptional set $\mathcal{N} \subset E$ such that $\mathbb{P}_x\{\Lambda\} = 1$ for $x \in E \backslash \mathcal{N}$ and $\Theta_t \Lambda \subset \Lambda$ for every $t \geq 0$;

(ii) $A_t|_\Lambda$ is measurable with respect to $\mathcal{F}_t|_\Lambda$ for each $t \geq 0$;

(iii) for $\omega \in \Lambda$, $A_t(\omega)$ is right-continuous and has left limits in $t \in [0, \infty)$ with $A_0(\omega) = 0$ and $|A_t(\omega)| < \infty$ for every $t < \infty$;

(iv) $A_{t+s}(\omega) = A_t(\omega) + A_s(\Theta_t \omega)$ for all $s, t \geq 0$.

If, in addition, the mapping $t \mapsto A_t(\omega)$ is positive and continuous for each $\omega \in \Lambda$, then A is called a *positive continuous additive functional* of X. We denote by $\mathcal{A}_c^+$ the family of all positive continuous additive functionals of

X. An additive functional admitting a defining set Λ with $\mathbb{P}_x\{\Lambda\} = 1$ for all $x \in E$ is called an *additive functional in the strict sense*.

The *energy* of an additive functional A of X is defined as

$$e(A) := \lim_{t \to 0+} \frac{1}{2t} \int_E \mathbb{E}_x A_t^2 \, dx$$

and an additive functional M of X is called a *martingale additive functional* if for each $t > 0$

$$\mathbb{E}_x M_t^2 < \infty, \quad \mathbb{E}_x M_t = 0 \quad \text{for q.e. } x \in E.$$

A martingale additive functional M is a martingale with respect to $\{\mathcal{F}_t, t \geq 0\}$ for q.e. $x \in E$. This follows directly from the additivity of M together with the Markov property of X

$$\mathbb{E}_x\{M_{t+s}|\mathcal{F}_s\} = \mathbb{E}_x\{M_s + M_t \circ \Theta_s | \mathcal{F}_s\} = M_s + \mathbb{E}_{X_s} M_t = M_s, \quad \mathbb{P}_x\text{-a.s.}$$

Two additive functionals $A^{(1)}$ and $A^{(2)}$ are *equivalent* if there are a common defining set Λ and a common exceptional set $\mathcal{N}$ such that for every $\omega \in \Lambda$ and q.e. $x \in E$

$$A_t^{(1)}(\omega) = A_t^{(2)}(\omega) \quad \mathbb{P}_x\text{-a.s. for all } t \geq 0.$$

We write $A^{(1)} \sim A^{(2)}$ in this case.

We can now formulate the *Revuz correspondence*: The family $\mathcal{A}_c^+/\sim$ of all equivalence classes of positive continuous additive functionals of X and the family $\mathcal{S}$ of all smooth measures with respect to the Dirichlet form $(\mathcal{E}, \mathcal{D}(\mathcal{E}))$ are in one-to-one correspondence. More precisely, for every $\mu_A \in \mathcal{S}$ there exists a unique (up to equivalence) $A \in \mathcal{A}_c^+$, and vice versa, such that

$$\lim_{t \to 0+} \frac{1}{t} \int_E \mathbb{E}_x \left\{ \int_0^t \phi(X_s) \, dA_s \right\} \psi(x) \, dx = \int_E \phi(x)\psi(x) \, d\mu_A(x)$$

for all non-negative Borel functions ϕ and all α-excessive functions ψ, cf. [54, Theorem 5.1.3]. μ_A is called the *Revuz measure* of the positive continuous additive functional A.

Definition A.26. The Hunt process X associated with a regular Dirichlet form $(\mathcal{E}, \mathcal{D}(\mathcal{E}))$ is called a *diffusion process* if

$$\mathbb{P}_x\{X_t \in C([0, \zeta); E)\} = 1, \quad \text{for all } x \in E\backslash\mathcal{N}.$$

If additionally

$$\mathbb{P}_x\{X_{\zeta-} \in E, \zeta < \infty\} = 0, \quad \text{for all } x \in E\backslash\mathcal{N},$$

then we say that X *admits no killing inside* E.

Numerous relations between the analytic and probabilistic concepts have been studied in the literature. In particular, one can derive the Beurling-Deny decomposition from Theorem A.14 in a purely probabilistic way, cf. [54, Chapter 5]. This yields the following probabilistic characterization of the local and the strongly local property.

Theorem A.27. *A regular Dirichlet form* $(\mathcal{E}, \mathcal{D}(\mathcal{E}))$ *is local if and only if* X *is equivalent to a diffusion process.* $(\mathcal{E}, \mathcal{D}(\mathcal{E}))$ *is strongly local if and only if* X *is equivalent to a diffusion process which admits no killing inside* E.

Let us conclude this section by recalling that in the framework of symmetric Dirichlet forms, the following *Fukushima decomposition* and the corresponding transformation formula, cf. [54, Chapter 5], play in some sense the roles of the Doob-Meyer decomposition and Itô's formula for semimartingales.

Theorem A.28. *Let* X *be the unique (up to equivalence) Hunt process associated with* $(\mathcal{E}, \mathcal{D}(\mathcal{E}))$ *and let* $v \in \mathcal{D}(\mathcal{E})$. *Then the composite process* $v(X) = (\Omega, \mathcal{F}, \{v(X_t), t \geq 0\}, \mathbb{P}_x)$ *admits the following unique decomposition for q.e.* $x \in E$

$$v(X_t) = v(X_0) + M_t^v + N_t^v \quad \text{for all } t > 0, \quad \mathbb{P}_x\text{-}a.s.,$$

where M^v *is a martingale additive functional of* X *having finite energy and* N^v *is a continuous additive functional of* X *having zero energy.*

Lemma A.29. *Let* $(\mathcal{E}, \mathcal{D}(\mathcal{E}))$ *be a regular Dirichlet form and suppose* $v \in \mathcal{D}(\mathcal{E})$ *satisfies*

$$\mathcal{E}(v, w) = \int_E w \, d\mu(x) \quad \text{for all } w \in \mathcal{D}(\mathcal{E}) \cap C_c(E),$$

where $\mu = \mu_1 - \mu_2$ *is a difference of positive Radon measures charging no sets of zero capacity. Then* $N^v = -A^{(1)} + A^{(2)}$, *where* $A^{(i)}$ *is the positive continuous additive functional with Revuz measure* μ_i, $i = 1, 2$.

By the Revuz correspondence, we may express the energy of M^v using its *predictable quadratic variation* $\langle M^v \rangle$. Namely, $\langle M^v \rangle \in \mathcal{A}_c^+$ satisfies

$$\mathbb{E}_x(M_t^v)^2 = \mathbb{E}_x\langle M^v \rangle_t, \quad \text{for all } t \geq 0 \text{ and q.e. } x \in E$$

and $\langle M^v \rangle$ is unique up to equivalence. Let $\mu_{\langle M^v \rangle}$ denote the Revuz measure of $\langle M^v \rangle$, then

$$e(M^v) = \lim_{t \to 0+} \frac{1}{2t} \int_D \mathbb{E}_x \langle M^v \rangle_t \, \mathrm{d}x = \frac{1}{2} \mu_{\langle M^v \rangle}(E).$$

The measure $\mu_{\langle M^v \rangle}$ is called the *energy measure* of M^v.

B Random field models

Let $D \subset \mathbb{R}^d$. A *random field* $\{Z(x,\omega),(x,\omega) \in D \times \Gamma\}$ is a family of real-valued random variables indexed by x on a probability space $(\Gamma, \mathcal{G}, \mathcal{P})$. That is, $Z : D \times \Gamma \to \mathbb{R}$ is a measurable mapping such that for each $x \in D$, $Z(x,\cdot) : \Gamma \to \mathbb{R}$ is a random variable. In this work, we are interested in *second order* random fields, i.e., fields such that $Z(x,\cdot)$ has finite variance. In this case, we can define the *mean* function $Z_0(x) := \mathbb{M}Z(x,\cdot)$ and the *covariance* function

$$\mathcal{C}_Z(x,y) := \mathbb{M}(Z(x,\cdot) - Z_0(x,\cdot))(Z(y,\cdot) - Z_0(y,\cdot)).$$

If $Z_0 = \text{const}$ and the covariance function depends only on the difference $x - y$, the random field is stationary and if the covariance function depends on $|x - y|$, it is *isotropic*.

A *Gaussian random field* $\{G(x,\omega),(x,\omega) \in D \times \Gamma\}$ is a second order random field, whose the finite-dimensional distributions $(G(x^{(1)}), ..., G(x^{(k)}))^T$ follow the multivariate Gaussian distribution for every finite collection $x^{(i)} \in D$, $i = 1, ..., k$. That is,

$$(G(x^{(1)}), ..., G(x^{(k)}))^T \sim \mathcal{N}((G_0(x^{(1)}), ..., G_0(x^{(k)}))^T, \mathcal{C}),$$

where the symmetric, non-negative definite matrix $\mathcal{C}$ has the entries

$$\mathcal{C}_{ij} := \mathcal{C}_G(x^{(i)}, x^{(j)}), \quad i, j = 1, ..., k.$$

In the geophysical literature, Gaussian random fields are commonly used with covariance function from the so-called *Matérn class*, that is,

$$\mathcal{C}_G^\gamma(x,y) := \sigma_G^2 \frac{2^{1-\gamma}}{\Gamma(\gamma)} \left(\frac{\sqrt{2\gamma}|x - y|}{\lambda} \right)^\gamma K_\gamma \left(\frac{\sqrt{2\gamma}|x - y|}{\lambda} \right), \quad x, y \in D \quad \text{(B.1)}$$

for positive standard deviation σ_G, correlation parameter λ and a positive parameter γ. K_γ is a modified Bessel function of the second kind of order γ, see [1]. Note that the regularity of the covariance function is controlled by the parameter γ. Moreover, for $\gamma = k + 1/2$, $k \in \mathbb{N}$, the covariance functions

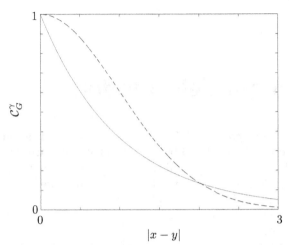

Figure B.1: Covariance functions (as functions of the distance $|x - y|$) for
$\gamma = 1/2$ (solid line) and $\gamma = \infty$ (dashed line).

become particularly simple, namely the product of an exponential term and
a polynomial of order k:

$$\sigma_G^2 \exp\left(- \frac{\sqrt{2\gamma}|x - y|}{\lambda} \right) \frac{\Gamma(k+1)}{\Gamma(2k+1)} \sum_{j=0}^{k} \frac{(k+j)!}{j!(k-j)!} \left(\frac{\sqrt{8\gamma}|x - y|}{\lambda} \right)^{k-j},$$

see [1]. Finally, notice that the choice $\gamma = 0.5$ yields the exponential
covariance function

$$\mathcal{C}_G^{1/2}(x, y) = \sigma_G^2 \exp\left(- \frac{|x - y|}{\lambda} \right)$$

and in the limit $\gamma \to \infty$, one obtains the squared exponential covariance
function

$$\mathcal{C}_G^{\infty}(x, y) = \sigma_G^2 \exp\left(- \frac{|x - y|^2}{2\lambda^2} \right).$$

Regularity of Gaussian random fields is commonly expressed in terms of
mean-square continuity and mean-square differentiability, which is intimately
related to the behavior of the covariance function at zero, cf. Potthoff [138].
We content ourselves here with stating the following continuity result.

Proposition B.1. *Let G denote a Gaussian random field with mean $G_0 \in C^{0,\delta}(\overline{D})$, $\delta \geq 0$ and covariance function (B.1) with γ such that, as a function of the distance $|x - y|$, C_G^γ belongs to $C^{0,1}(\mathbb{R}^+)$. Then the log-normal random field*

$$\{\kappa(x,\omega) := \exp(G(x,\omega)), (x,\omega) \in \overline{D} \times \Gamma\}$$

satisfies $\kappa \in L^2(\Gamma; C^{0,\delta}(\overline{D}))$, $\delta < 1/2$.

The circulant embedding simulation technique

For the simulation of two-dimensional stationary Gaussian random fields, we use the following *circulant embedding* technique introduced by Dietrich and Newsam [41]: Recall that an $M \times M$ matrix B is called *Toeplitz* if the entries along each diagonal are the same. A Toeplitz matrix for which each column is a circular shift of the elements in the preceding column is called a *circulant* matrix. Note that symmetric Toeplitz matrices can always be extended to give symmetric circulant matrices by padding them with additional rows and columns. A matrix is called block Toeplitz with Toeplitz blocks (BTTB) if each row of blocks is a periodic shift of its previous row of blocks and every block is a Toeplitz matrix. Analogously, a matrix is called block circulant with circulant blocks (BCCB) if each row of blocks is a periodic shift of its previous row of blocks and every block is a circulant matrix. Notice that symmetric BTTB matrices can always be extended to symmetric BCCB matrices. Note that every BCCB matrix

$$B = \begin{pmatrix} B_0 & B_{n-1} & \cdots & B_2 & B_1 \\ B_1 & B_0 & B_{n-1} & \ddots & B_2 \\ \vdots & \ddots & \ddots & \ddots & \vdots \\ B_{n-2} & \ddots & B_1 & B_0 & B_{n-1} \\ B_{n-1} & B_{n-2} & \cdots & B_1 & B_0 \end{pmatrix},$$

where each block B_k is an $n \times n$ circulant matrix, may be stored in the handy *reduced* format

$$B_{\mathrm{red}} = (b_0, ..., b_{n-1}) \in \mathbb{R}^{n \times n},$$

where $b_k \in \mathbb{R}^n$ denotes the first column of B_k, $k = 0, ..., n - 1$.

With regard to numerics, BCCB matrices are convenient as their eigenstructure is well known, namely an $n^2 \times n^2$ BCCB matrix B admits the decomposition

$$B = P\Lambda P^*,$$

Algorithm 7 Independent $\mathcal{N}(0, B)$-distributed realizations, where B is a $n^2 \times n^2$ BCCB covariance matrix

---

Require: Reduced matrix B_{red}
    Calculate $\Lambda \leftarrow n^2 \mathtt{ifft2}(B_{\mathrm{red}})$
    Draw $\gamma_1, \gamma_2 \sim \mathcal{N}(0, I_n)$
    Set $\gamma \leftarrow \gamma_1 + i\gamma_2$
    Calculate $V \leftarrow (\Lambda.\hat{\ }0.5). * \gamma$
    Calculate $Z = \mathtt{fft2}(V)/n$
    return $\mathrm{Re}(Z), \mathrm{Im}(Z)$

---

where Λ denotes the diagonal matrix of eigenvalues and P is the Kronecker product of two discrete Fourier transform matrices, i.e., $P = F \otimes F$ with $F_{jk} = \exp(-2\pi ijk/n)/\sqrt{n}$, $j, k = 0, 1, ..., n-1$.

Suppose we want to draw realizations from $\mathcal{N}(0, B)$. For B to be a proper covariance matrix, that is, symmetric and non-negative definite, it must in particular have real, non-negative eigenvalues. That is, we can define the component-wise square root $\Lambda^{1/2}$ and consider

$$Z = P\Lambda^{1/2}\gamma, \quad \gamma \sim \mathcal{N}_c(0, 2I_{n^2}),$$

where $\mathcal{N}_c$ denotes the complex Gaussian distribution. Then it holds that $Z \sim \mathcal{N}_c(0, 2B)$ so that $\mathrm{Re}(Z)$ and $\mathrm{Im}(Z)$ are independent $\mathcal{N}(0, B)$-distributed realizations. Multiplication with P can be done efficiently via FFT and Λ can be obtained via inverse FFT, see Algorithm 7.

In general, the covariance matrix $\mathcal{C}$ of a two-dimensional Gaussian random field is not a BCCB matrix. However, given a stationary Gaussian random field and a uniformly spaced discretization dividing D into $n \times n$ squares, where the sample points $x^{(i)}$ are the lexicographically ordered mid-points of these squares, $\mathcal{C}$ is a BTTB matrix with $n \times n$ Toeplitz blocks each of size $n \times n$. Since any such matrix can be embedded into a larger BCCB matrix B, realizations drawn from $\mathcal{N}(0, \mathcal{C})$ can be obtained from realizations drawn from $\mathcal{N}(0, B)$ by discarding some of the sample points. However, the extended BCCB is not necessarily a proper covariance matrix, indeed it may fail to be non-negative definite. Therefore, we have to check the eigenvalues and if there are negative ones, then a larger BCCB matrix must be constructed. Figure B.2 shows a BTTB covariance matrix and its embedding into a larger BCCB matrix.

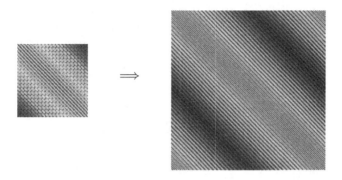

Figure B.2: `imagesc` plots of a BTTB covariance matrix $\mathcal{C}$ with 25×25 Toeplitz blocks and its embedding into a larger BCCB matrix.

If this procedure becomes intractable because of the computational cost, one usually uses an approximation. More precisely, we may decompose the BCCB matrix B in the form

$$B = P\Lambda_+ P^* + P\Lambda_- P^*,$$

where the diagonal matrix Λ_+ contains the positive eigenvalues of B and Λ_1 the negative ones. Then using $\widehat{B} := P\Lambda_+ P^*$ instead of B yields an approximation error which may be controlled by the size of the neglected negative eigenvalues. For a more detailed study of the circulant embedding technique we refer to the monograph [114].

C FEM discretization of the forward problem

In order to solve the EIT forward problem for the complete electrode model numerically, we have to compute both, the potential and the voltage patterns, given the contact impedance, the injected current patterns and the conductivity. For simplicity of the presentation, we assume here that $d = 2$, $|E_l| = |E|$, $l = 1, ..., N$, that the function z is constant and that the conductivity κ is isotropic. For the numerical solution of the forward problem, we used a finite element (FEM) discretization of (1.3), (1.11), (1.12), which has been discussed previously, e.g., in [77, 92].

The FEM approximation is based on the variational formulation given in the article [150], where it was shown that there exists a unique weak solution $(u, U) \in H^1(D) \oplus \mathbb{R}^N$ satisfying the grounding condition (1.10) such that for all $(v, V) \in H^1(D) \oplus \mathbb{R}^N$

$$\int_D \kappa \nabla u \cdot \nabla v \, dx + \sum_{l=1}^{N} \frac{1}{z} \int_{E_l} (u - U_l)(v - V_l) \, d\sigma(x) = \sum_{l=1}^{N} J_l V_l. \quad \text{(C.1)}$$

Assume that the domain D may be divided into a mesh

$$\mathcal{T}_h = \{T_1, ..., T_{m_h}\}$$

of m_h disjoint triangles with maximal edge length h, such that $D = \cup_{k=1}^{m_h} T_k$, joined at n_h vertex nodes. Given a current pattern J, let $u^h \in Q^{\mathcal{T}_h} \subset H^1(D)$ denote the finite dimensional approximation for the solution u from the ansatz space $Q^{\mathcal{T}_h} := \text{span}\{\phi_1, ..., \phi_{n_h}\}$, where the functions ϕ_j, $j = 1, ..., n_h$, are the nodal basis functions of the triangulation $\mathcal{T}_h$. In this work, we use standard piecewise linear basis functions. The approximation is given by (u^h, U^h) with

$$u^h = \sum_{j=1}^{n_h} \alpha_j \phi_j, \quad U^h = \sum_{l=1}^{N-1} \beta_l(e_1 - e_{l+1}),$$

where the vectors e_l, $l = 1, ..., N$, are the Cartesian basis vectors of $\mathbb{R}^N$ which enforces the grounding condition (1.10). Substituting these approximations into the variational formulation (C.1) yields an $(n_h + N - 1)$-dimensional linear system

$$Ax = \begin{pmatrix} \mathbf{0} \\ \mathcal{D}^T J \end{pmatrix},$$

where $\mathbf{0} = (0, ..., 0)^T \in \mathbb{R}^{n_h}$ and

$$\mathcal{D} = \begin{pmatrix} 1 & 1 & 1 & \cdots & 1 \\ -1 & 0 & 0 & \cdots & 0 \\ 0 & -1 & 0 & \cdots & 0 \\ \vdots & & \ddots & & \vdots \\ 0 & \cdots & 0 & -1 & 0 \\ 0 & \cdots & 0 & 0 & -1 \end{pmatrix} \in \mathbb{R}^{N \times (N-1)}.$$

The stiffness matrix $A \in \mathbb{R}^{(n_h+N-1)\times(n_h+N-1)}$ is a sparse, symmetric and positive definite block matrix of the form

$$A = \begin{pmatrix} B & C \\ C^T & G \end{pmatrix}$$

with blocks

$$B = \left(\int_D \kappa \nabla \phi_i \cdot \nabla \phi_j \, \mathrm{d}x + \sum_{l=1}^N \frac{1}{z} \int_{E_l} \phi_i \phi_j \, \mathrm{d}\sigma(x) \right)_{1 \leq i,j \leq n_h}$$

$$C = \left(-\frac{1}{z} \left(\int_{E_1} \phi_i \, \mathrm{d}\sigma(x) - \int_{E_{j+1}} \phi_j \, \mathrm{d}\sigma(x) \right) \right)_{1 \leq i \leq n_h, \ 1 \leq j \leq N-1}$$

$$G = \left(\sum_{l=1}^N \frac{1}{z} \int_{E_l} (e_1 - e_{i+1})_l (e_1 - e_{j+1})_l \, \mathrm{d}\sigma(x) \right)_{1 \leq i,j \leq N-1}$$

$$= \frac{|E|}{z} I_{N-1} + \left(\frac{|E|}{z} \right)_{1 \leq i,j \leq N-1}.$$

The solution vector is given by $x = (\alpha, \beta)^T \in \mathbb{R}^{n_h+N-1}$ and the voltage vector is calculated as $U^h = \mathcal{D}\beta$. Note that we may write equivalently

$$U^h = \mathcal{D}(A^{-1})_{n_h+1 \leq i,j \leq n_h+N-1} \mathcal{D}^T J =: R^h_{z,\kappa} J,$$

where the matrix $R^h_{z,\kappa} \in \mathbb{R}^{N \times N}$ is the *discrete current-to-voltage map* based on the described FEM discretization with maximal edge length h.

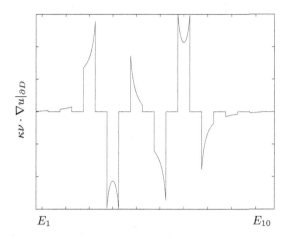

Figure C.1: FEM approximation of the boundary current density $\kappa\nu \cdot \nabla u|_{\partial D}$; x-axis ticks correspond to the midpoints of the electrodes E_l, $l = 1, ..., 10$.

For a more advanced hp-adaptive FEM solution to the complete electrode model forward problem, we refer the reader to the article [65] by Hakula, Hyvönen and Tuominen.

Note that the Robin boundary condition (1.12) in the complete electrode model leads to singular behavior of the boundary current density at the end points of the electrodes, see Figure C.1. As a result of this behavior, numerical approximations, both via finite element and boundary element methods, require very fine discretization in a vicinity of the electrodes. In fact, the numerical difficulties become more severe as the contact impedance decreases and the shunt model is approached. This is somewhat unfortunate since in practice, one typically aims for a small contact impedance which corresponds to good electrical contacts.

Bibliography

[1] M. Abramowitz and I. A. Stegun. *Handbook of mathematical functions with formulas, graphs, and mathematical tables*, National Bureau of Standards Applied Mathematics Series, U.S. Government Printing Office, Washington, D.C., 1964.

[2] K. Akhlil. *Probabilistic solution of the general Robin boundary value problem on arbitrary domains*, Int. J. Stoch. Anal., (2012), Art. ID 163096, 17.

[3] G. Alessandrini. *Stable determination of conductivity by boundary measurements*, Appl. Analysis, **27** (1988), 153–172.

[4] A. Alexanderian, M. Rathinam and R. Rostamian. *Homogenization, symmetry, and periodization in diffusive random media*, Acta Math. Sci. Ser. B Engl. Ed., **32** (2012), 129–154.

[5] W. Arendt and N. Nikolski. *Vector-valued holomorphic functions revisited*, Math. Z., **234** (2000), 777–805.

[6] S. R. Arridge, J. P. Kaipio, V. Kolehmainen, M. Schweiger, E. Somersalo, T. Tarvainen and M. Vauhkonen. *Approximation errors and model reduction with an application in optical diffusion tomography*, Inverse Problems, **22** (2006), 175–195.

[7] K. Astala and L. Päivärinta. *Calderón's inverse conductivity problem in the plane*, Ann. of Math., **163** (2006), 265–299.

[8] K. Astala, L. Päivärinta and M. Lassas. *Calderón's inverse problem for anisotropic conductivity in the plane*, Comm. Partial Differential Equations, **30** (2005), 207–224.

[9] M. Avellaneda, Th. Y. Hou and G. C. Papanicolaou. *Finite difference approximations for partial differential equations with rapidly oscillating coefficients*, RAIRO Modél. Math. Anal. Numér., **25** (1991), 693–710.

[10] I. Babuska, R. Tempone and G. E. Zouraris. *Solving elliptic boundary value problems with uncertain coefficients by the finite element method: the stochastic formulation*, Comput. Methods Appl. Mech. Engrg., **194** (2005), 1251–1294.

[11] H. T. Banks and A. K. Criner. *Thermal based methods for damage detection and characterization in porous materials*, Inverse Problems, **28** (2012), 065021.

[12] J. A. Barceló, T. Barceló and A. Ruiz. *Stability of the inverse conductivity problem in the plane for less regular conductivities*, J. Differential Equations, **173** (2001), 231–270.

[13] G. Barles, F. Da Lio, P.-L. Lions and P. E. Souganidis. *Ergodic problems and periodic homogenization for fully non-linear equations in half-space type domains with Neumann boundary conditions*, Indiana Univ. Math. J., **57** (2008), 2355–2376.

[14] R. F. Bass. *Uniqueness for the Skorokhod equation with normal reflection in Lipschitz domains*, Electron. J. Probab., **1** (1996), 1083-6489.

[15] R. F. Bass and P. Hsu. *Some potential theory for reflecting Brownian motion in Hölder and Lipschitz domains*, Ann. Probab., **19** (1991), 486–508.

[16] R. F. Bass and P. Hsu. *Pathwise uniqueness for reflecting Brownian motion in Euclidean domains*, Probab. Theory Related Fields, **117** (2000), 183–200.

[17] M. Bebendorf. *Hierarchical matrices: A means to efficiently solve elliptic boundary value problems*, Springer-Verlag, Berlin, 2008.

[18] A. Benchérif-Madani and É. Pardoux. *A probabilistic formula for a Poisson equation with Neumann boundary condition*, Stoch. Anal. Appl., **27** (2009), 739–746.

[19] A. Bensoussan, J. L. Lions and G. Papanicolaou. *Asymptotic analysis for periodic structures*, AMS Chelsea Publishing, Providence, RI, 1978.

[20] A. Beurling and J. Deny. *Dirichlet spaces*, Proc. Nat. Acad. Sci. U.S.A., **45** (1959), 208–215.

[21] P. Billingsley. *Probability and measure*, John Wiley & Sons Inc., New York, 1995.

[22] S. Börm, L. Grasedyck and W. Hackbusch. *Introduction to hierarchical matrices with applications*, Engineering Analysis with Boundary Elements, **27** (2003), 405–422.

[23] L. Borcea. *Electrical impedance tomography*, Inverse Problems, **18** (2002), R99–R136.

[24] A. N. Borodin and P. Salminen. *Handbook of Brownian motion - facts and formulae*, Birkhauser Verlag, Basel-Boston-Berlin, 2002.

[25] M. Bossy, N. Champagnat, S. Maire and D. Talay. *Probabilistic interpretation and random walk on spheres algorithms for the Poisson-Boltzmann equation in molecular dynamics*, M2AN Math. Model. Numer. Anal., **44** (2010), 997–1048.

[26] A. Bourgeat and A. Piatnitski. *Approximations of effective coefficients in stochastic homogenization*, Ann. Inst. H. Poincaré Probab. Statist., **40** (2004), 153–165.

[27] G. A. Brosamler. *A probabilistic solution of the Neumann problem*, Math. Scand., **38** (1976), 137–147.

[28] A. Calderón. *On an inverse boundary value problem*, in *Seminar on Numerical Analysis and its Applications to Continuum Physics*, Soc. Brasileira de Matematica, Rio de Janeiro, 65–73, 1980.

[29] E. A. Carlen, S. Kusuoka and D. W. Stroock. *Upper bounds for symmetric Markov transition functions*, Ann. Inst. H. Poincaré Probab. Statist., **23** (1987), 245–287.

[30] P. Caro and K. Rogers. *Global uniqueness for the Calderón problem with Lipschitz conductivities*, Preprint (2014), arXiv:1411.8001.

[31] J. Charrier. *Strong and weak error estimates for the solutions of elliptic partial differential equations with random coefficients*, SIAM J. Numer. Anal, **50** (2012), 216–246.

[32] Z. Q. Chen. *On reflecting diffusion processes and Skorokhod decompositions*, Probab. Theory Related Fields, **94** (1993), 281–315.

[33] Z. Q. Chen, D. A. Croydon and T. Kumagai. *Quenched invariance principles for random walks and elliptic diffusions in random media with boundary*, to appear in Ann. Probab.

[34] Z. Q. Chen and T. Zhang. *Time-reversal and elliptic boundary value problems*, Ann. Probab., **37** (2009), 1008–1043.

[35] Z. Q. Chen and T. Zhang. *A probabilistic approach to mixed boundary value problems for elliptic operators with singular coefficients*, Proc. Amer. Math. Soc., **142** (2014), 2135–2149.

[36] M. Cheney, D. Isaacson and J. C. Newell. *Electrical impedance tomography*, SIAM Rev., **41** (1999), 85–10.

[37] R. Costaouec, C. Le Bris and F. Legoll. *Variance reduction in stochastic homogenization: proof of concept, using antithetic variables*, Bol. Soc. Esp. Mat. Apl. SēMA, **50** (2010), 9–26.

[38] M. Deaconu and A. Lejay. *A random walk on rectangles algorithm*, Methodol. Comput. Appl. Probab., **8** (2006), 135–151.

[39] A. De Masi, P. A. Ferrari, S. Goldstein and W. D. Wick. *An invariance principle for reversible Markov processes. Applications to random motions in random environments*, J. Stat. Phys., **55** (1988), 787–855.

[40] A. De Wit. *Correlation structure dependence of the effective permeability of heterogeneous porous media*, Phys. Fluids, **7** (1995), 2553–2562.

[41] C. R. Dietrich and G. N. Newsam. *Fast and exact simulation of stationary Gaussian processes through circulant embedding of the covariance matrix*, SIAM J. Sci. Comput., **18** (1997), 1088–1107.

[42] D. Dos Santos Ferreira, C. Kenig, M. Salo and G. Uhlmann. *Limiting Carleman weights and anisotropic inverse problems*, Invent. Math., **178** (2009), 119–171.

[43] A.-C. Egloffe, A. Gloria., J.-C. Mourrat and T. N. Nguyen. *Random walk in random environment, corrector equation, and homogenized coefficients: from theory to numerics, back and forth*, IMA J. Num. Anal., (2014), dru010.

[44] H. W. Engl, M. Hanke and A. Neubauer. *Regularization of inverse problems*, Kluwer Academic Publishers Group, Dordrecht, 1996.

[45] P. Étoré. *On random walk simulation of one-dimensional diffusion processes with discontinuous coefficients*, Electron. J. Probab., **11** (2006), no. 9, 249–275 (electronic).

[46] P. Étoré and A. Lejay. *A Donsker theorem to simulate one-dimensional processes with measurable coefficients*, ESAIM Probab. Stat., **11** (2007), 301–326 (electronic).

[47] P. Étoré and M. Martinez. *Exact simulation of one-dimensional stochastic differential equations involving the local time at zero of the unknown process*, Monte Carlo Methods Appl., **19** (2013), 41–71.

[48] R. P. Feynman. *The principle of least action in quantum mechanics*, Ph.D thesis, Princeton University, Princeton, NJ, 1942.

[49] R. Figari, E. Orlandi and G. Papanicolaou. *Mean field and Gaussian approximation for partial differential equations with random coefficients*, SIAM J. Appl. Math., **42** (1982), 1069–1077.

[50] P. J. Fitzsimmons. *Even and odd continuous additive functionals*, in *Dirichlet forms and stochastic processes (Beijing, 1993)*, de Gruyter, Berlin (1995), 139–154.

[51] M. Freidlin. *Functional integration and partial differential equations*, Princeton University Press, Princeton, NJ, 1985.

[52] M. Fukushima. *Dirichlet spaces and strong Markov processes*, Trans. Amer. Math. Soc., **162** (1971), 185–224.

[53] M. Fukushima. *On a decomposition of additive functionals in the strict sense for a symmetric Markov process*, in *Dirichlet forms and stochastic processes (Beijing, 1993)*, de Gruyter, Berlin, 155–169, 1995.

[54] M. Fukushima, Y. Ōshima and M. Takeda. *Dirichlet forms and symmetric Markov processes*, Walter de Gruyter & Co., Berlin, 1994.

[55] M. Fukushima and M. Tomisaki. *Construction and decomposition of reflecting diffusions on Lipschitz domains with Hölder cusps*, Probab. Theory Related Fields, **106** (1996), 521–557.

[56] A. Gloria and J.-C. Mourrat. *Quantitative version of the Kipnis-Varadhan theorem and Monte Carlo approximation of homogenized coefficients*, Ann. Appl. Probab., **23** (2013), 1544–1584.

[57] A. Gloria, S. Neukamm and F. Otto. *Quantification of ergodicity in stochastic homogenization: optimal bounds via spectral gap on Glauber dynamics*, Invent. Math. (2014), DOI 10.1007/s00222-014- 0518-z.

[58] A. Gloria and F. Otto. *Quantitative results on the corrector equation in stochastic homogenization*, Preprint (2014), arXiv:1409.0801.

[59] L. Grasedyck, R. Kriemann and S. Le Borne. *Parallel blackbox H-LU preconditioning for elliptic boundary value problems*, Computing and Visualization in Sciences, **11** (2007), 273–291.

[60] J. A. Griepentrog and L. Recke. *Linear elliptic boundary value problems with non-smooth data: normal solvability on Sobolev-Campanato spaces*, Math. Nachr., **225** (2001), 39–74.

[61] P. Grisvard. *Elliptic problems in nonsmooth domains*, Pitman, Boston, MA, 1985.

[62] B. Haberman and D. Tataru. *Uniqueness in Calderón's problem with Lipschitz conductivities*, Duke Math. J., **162** (2013), 496–516.

[63] B. Haberman *Uniqueness in Calderón's problem for conductivities with unbounded gradient*, Preprint (2014), arXiv:1410.2201.

[64] H. Hakula, N. Hyvönen and M. Leinonen. *Reconstruction algorithm based on stochastic Galerkin finite element method for electrical impedance tomography*, Inverse Problems, **30** (2014), 065006.

[65] H. Hakula, N. Hyvönen and T. Tuominen. *On the hp-adaptive solution of complete electrode model forward problems of electrical impedance tomography*, J. Comput. Appl. Math., **236** (2012), 4645–4659.

[66] M. Hanke, N. Hyvönen, S. Reusswig. *An inverse backscatter problem for electric impedance tomography*, SIAM J. Math. Anal., **41** (2009), 1948–1966.

[67] M. Hanke, N. Hyvönen, S. Reusswig. *Convex backscattering support in electric impedance tomography*, Numer. Math., **117** (2011), 373–396.

[68] M. Hanke and B. Schappel. *The factorization method for electrical impedance tomography in the half-space*, SIAM J. Appl. Math., **68** (2008), 907–924.

[69] P. Hsu. *Probabilistic approach to the Neumann problem*, Comm. Pure Appl. Math., **38** (1985), 445–472.

[70] P. Hsu. *On the Poisson kernel for the Neumann problem of Schrödinger operators*, J. London Math. Soc. (2), **36** (1987), 370–384.

[71] N. Hyvönen. *Comparison of idealized and electrode Dirichlet-to-Neumann maps in electric impedance tomography with an application to boundary determination of conductivity*, Inverse Problems, **25** (2009), 085008.

[72] N. Ikeda and S. Watanabe. *Stochastic differential equations and diffusion processes*, North-Holland Publishing Co., Amsterdam-New York; Kodansha, Ltd., Tokyo, 1981.

[73] V. Isakov. *On uniqueness of recovery of a discontinuous conductivity coefficient*, Comm. Pure Appl. Math., **41** (1988), 865–877.

[74] K. Itô and H. P. McKean Jr. *Diffusion processes and their sample paths*, Springer-Verlag, Berlin-New York, 1974.

[75] I. Jankovic, A. Fiori and G. Dagan. *Effective conductivity of an isotropic heterogeneous medium of lognormal conductivity distribution*, Multiscale Model. Simul., **1** (2003), 40–56 (electronic).

[76] M. Kac. *On distributions of certain Wiener functionals*, Trans. Amer. Math. Soc., **65** (1949), 1–13.

[77] J. P. Kaipio, V. Kolehmainen, E. Somersalo and M. Vauhkonen. *Statistical inversion and Monte Carlo sampling methods in electrical impedance tomography*, Inverse Problems, **16** (2000), 1487–1522.

[78] J. Kaipio and E. Somersalo. *Statistical and computational inverse problems*, Springer-Verlag, New York, 2005.

[79] J. Kaipio and E. Somersalo. *Statistical inverse problems: discretization, model reduction and inverse crimes*, J. Comput. Appl. Math., **198** (2007), 493–504.

[80] I. Karatzas and S. E. Shreve. *Brownian motion and stochastic calculus*, Springer-Verlag, New York, 1991.

[81] M. R. Karim and K. Krabbenhoft. *New renormalization schemes for conductivity upscaling in heterogeneous media*, Transp. Porous Media, **85** (2010), 677–690.

[82] R. S. Keskin, M. D. Grigoriu. *A probability-based method for calculating effective diffusivity coefficients of composite media* Probabilistic Engineering Mechanics, **25** (2010), 249–254.

[83] I. C. Kim, D. Cule and S. Torquato. *Comment on "Walker diffusion method for calculation of transport properties of composite materials"*, Phys. Rev. E, **61** (2000), 4659–4660.

[84] I. C. Kim and S. Torquato. *First passage time calculation of the conductivity of continuum models of multiphase composites*, Physical Review A, **43** (1991), 3198–3201.

[85] I. C. Kim and S. Torquato. *Effective conductivity of suspensions of spheres by Brownian motion simulation* Journal of Applied Physics, **69** (1991), 2280–2289.

[86] C. Kipnis and S. R. S. Varadhan. *Central limit theorem for additive functional of reversible Markov processes and applications to simple exclusion*, Commun. Math. Phys., **104** (1986), 1–19.

[87] A. Kirsch. *An introduction to the mathematical theory of inverse problems*, Springer-Verlag, New York, 2011.

[88] A. Klenke. *Probability theory*, Springer, London, 2014.

[89] P. E. Kloeden and E. Platen. *Numerical solution of stochastic differential equations*, Springer-Verlag, Berlin, 1992.

[90] K. Knudsen, M. Lassas, J. Mueller and S. Siltanen. *Regularized D-bar method for the inverse conductivity problem*, Inverse Probl. Imaging, **3** (2009), 599–624.

[91] R. Kohn and M. Vogelius. *Determining conductivity by boundary measurements*, Comm. Pure Appl. Math., **37** (1984), 289–298.

[92] V. Kolehmainen. *Novel approaches to image reconstruction in diffusion tomography*, Ph.D. thesis, Kuopio University Publications, Koupio, Finland, 2001.

[93] V. Kolehmainen, M. Lassas and P. Ola. *Electrical impedance tomography problem with inaccurately known boundary and contact impedances*, IEEE Transactions on Medical Imaging., **27** (2008), 1404–1414.

[94] C. Kronsbein. *On selected efficient numerical methods for multiscale problems with stochastic coefficients*, Ph.D thesis, University of Kaiserslautern, Kaiserslautern, Germany, 2012.

[95] O. A. Ladyzenskaya and N. N. Ural′ceva. *Linear and quasilinear elliptic equations*, Academic Press, New York, 1968.

[96] S. Lasanen. *Discretizations of generalized random variables with applications to inverse problems*, Ann. Acad. Sci. Fenn. Math. Diss. No. 130, 2002.

[97] S. Lasanen. *Non-Gaussian statistical inverse problems. Part I: Posterior distributions*, Inverse Probl. Imaging, **6** (2012), 215–266.

[98] S. Lasanen. *Non-Gaussian statistical inverse problems. Part II: Posterior convergence for approximated unknowns.*, Inverse Probl. Imaging, **6** (2012), 267–287.

[99] P. L'Ecuyer. *Random number generation*, in *Handbook of computational statistics—concepts and methods*, Springer, Heidelberg (2012), 35–71.

[100] A. Lehikoinen, S. Finsterle, A. Voutilainen, L. M. Heikkinen, M. Vauhkonen and J. P. Kaipio. *Approximation errors and truncation of computational domains with application to geophysical tomography*, Inverse Probl. Imaging, **1** (2007), 371–389.

[101] M. S. Lehtinen, L. Päivärinta and E. Somersalo. *Linear inverse problems for generalised random variables*, Inverse Problems **5** (1989), 599–612.

[102] M. Leinonen, H. Hakula, and N. Hyvönen. *Application of stochastic Galerkin FEM to the complete electrode model of electrical impedance tomography*, J. Comput. Phys., **269** (2014), 181–200.

[103] A. Lejay. *A probabilistic approach to the homogenization of divergence-form operators in periodic media*, Asymptot. Anal., **28** (2001), 151–162.

[104] A. Lejay. *Homogenization of divergence-form operators with lower-order terms in random media*, Probab. Theory Related Fields, **120** (2001), 255–276.

[105] A. Lejay. *On the constructions of the skew Brownian motion*, Probab. Surv., **3** (2006), 413–466.

[106] A. Lejay. *Simulation of a stochastic process in a discontinuous layered medium*, Electron. Commun. Probab. **16** (2011), 764–774.

[107] A. Lejay and S. Maire. *Simulating diffusions with piecewise constant coefficients using a kinetic approximation*, Comput. Methods Appl. Mech. Engrg., **199** (2010), 2014–2023.

[108] A. Lejay and S. Maire. *New Monte Carlo schemes for simulating diffusions in discontinuous media*, J. Comput. Appl. Math., **245** (2013), 97–116.

[109] A. Lejay and M. Martinez. *A scheme for simulating one-dimensional diffusion processes with discontinuous coefficients*, Ann. Appl. Probab., **16** (2006), 107–139.

[110] A. Lejay and G. Pichot. *Simulating diffusion processes in discontinuous media: a numerical scheme with constant time steps*, J. Comput. Phys., **231** (2012), 7299–7314.

[111] D. Lépingle. *Euler scheme for reflected stochastic differential equations*, Math. Comput. Simulation, **38** (1995), 119–126.

[112] P.-L. Lions and A.-S. Sznitman. *Stochastic differential equations with reflecting boundary conditions*, Comm. Pure Appl. Math., **37** (1984), 511–537.

[113] J. S. Liu. *Monte Carlo strategies in scientific computing*, Springer, New York, 2008.

[114] G. J. Lord, C. E. Powell and T. Shardlow. *An introduction to computational stochastic PDEs*, Cambridge University Press, 2014.

[115] Z. M. Ma and M. Röckner. *Introduction to the theory of (nonsymmetric) Dirichlet forms*, Springer-Verlag, Berlin, 1992.

[116] S. Maire and G. Nguyen, *Stochastic finite differences for elliptic diffusion equations in stratified domains*, Preprint (2013), hal-00809203.

[117] S. Maire and M. Simon. *A partially reflecting random walk on spheres algorithm for electrical impedance tomography*, Preprint (2015), arXiv:1502.04318.

[118] N. Mandache. *Exponential instability in an inverse problem for the Schrödinger equation*, Inverse Problems, **17** (2001), 1435–1444.

[119] A. Mantoglou and J. L. Wilson. *The turning bands method for simulation of random fields using line generation by a spectral method*. Water Resour. Res. **18** (1982), 1379–1394.

[120] M. Martinez and D. Talay. *One-dimensional parabolic diffraction equations: pointwise estimates and discretization of related stochastic differential equations with weighted local times*, Electron. J. Probab., **17** (2012), no. 27, 30.

[121] M. Mascagni and C. Hwang. *ε-shell error analysis for "walk on spheres" algorithms*, Math. Comput. Simulation, **63** (2003), 93–104.

[122] M. Mascagni and N. A. Simonov. *Monte Carlo methods for calculating some physical properties of large molecules*, SIAM J. Sci. Comput., **26** (2004), 339–357.

[123] G. N. Milstein and M. V. Tretyakov. *Simulation of a space-time bounded diffusion*, Ann. Appl. Probab., **9** (1999), 732–779.

[124] J.-C. Mourrat. *Variance decay for functionals of the environment viewed by the particle*, Ann. Inst. Henri Poincaré Probab. Stat., **47** (2011), 294–327.

[125] M. E. Muller. *Some continuous Monte Carlo methods for the Dirichlet problem*, Ann. Math. Statist., **27** (1956), 569–589.

[126] A. I. Nachman. *Global uniqueness for a two-dimensional inverse boundary value problem*, Ann. Math., **143** (1996), 71–96.

[127] J. Nash. *Continuity of solutions of parabolic and elliptic equations*, Amer. J. Math., **80** (1958), 931–954.

[128] A. Nissinen, V. Kolehmainen and J. P. Kaipio. *Reconstruction of domain boundary and conductivity in electrical impedance tomography using the approximation error approach*, Int. J. Uncertain. Quantif., **1** (2011), 203–222.

[129] R. Nittka. *Regularity of solutions of linear second order elliptic and parabolic boundary value problems on Lipschitz domains*, J. Differential Equations, **251** (2011), 860–880.

[130] J. Nolen and G. Papanicolaou. *Fine scale uncertainty in parameter estimation for elliptic equations*, Inverse Problems, **25** (2009), 115021, 22.

[131] G. C. Papanicolaou. *Diffusion in random media*, Surveys Appl. Math., **1** (1995), 205–253.

[132] G. C. Papanicolaou and S. R. S. Varadhan. *Boundary value problems with rapidly oscillating random coefficients*, in Random fields, Vol. I, II (Esztergom, 1979), 835–873, North-Holland, Amsterdam, 1981.

[133] V. G. Papanicolaou. *The probabilistic solution of the third boundary value problem for second order elliptic equations*, Probab. Theory Related Fields, **87** (1990), 27–77.

[134] A. Pazy. *Semigroups of linear operators and applications to partial differential equations*, Springer-Verlag, New York, 1983.

[135] A. Piatnitski and E. Remy. *Homogenization of elliptic difference operators*, SIAM J. Math. Anal., **33** (2001), 53–83.

[136] P. Piiroinen. *Statistical measurements, experiments and applications.* Ann. Acad. Sci. Fenn. Math. Diss. No. 143, 2005.

[137] P. Piiroinen and M. Simon. *From Feynman-Kac formulae to numerical stochastic homogenization in electrical impedance tomography*, Preprint (2015), arXiv:1502.04353.

[138] J. Potthoff. *Sample properties of random fields III: differentiability*, Comm. Stoch. Anal., **4** (2010), 335–353.

[139] S. Pursiainen. *Two-stage reconstruction of a circular anomaly in electrical impedance tomography*, Inverse Problems, **22** (2006), 1689–1703.

[140] D. Revuz and M. Yor. *Continuous martingales and Brownian motion*, Springer-Verlag, Berlin, 1994.

[141] R. Rhodes. *Stochastic homogenization of reflected stochastic differential equations*, Electron. J. Probab., **15** (2010), 989–1021.

[142] L. Roininen, M. S. Lehtinen, S. Lasanen, M. Orispää and M. Markkanen. *Correlation priors*, **5** (2011), 167–184.

[143] L. Roininen, J. Huttunen and S. Lasanen. *Whittle-Matérn priors for Bayesian statistical inversion with applications in electrical impedance tomography*, Inverse Probl. Imaging, **8** (2014), 561–586.

[144] L. Roininen, P. Piiroinen and M. Lehtinen. *Constructing continuous stationary covariances as limits of the second-order stochastic difference equations*, Inverse Probl. Imaging, **7** (2013), 611–647.

[145] A. Rozkosz. *On a decomposition of symmetric diffusions with reflecting boundary conditions*, Stochastic Process. Appl., **103** (2003), 101–122.

[146] A. Rozkosz and L. Słomiński. *Stochastic representation of reflecting diffusions corresponding to divergence form operators*, Studia Math., **139** (2000), 141–174.

[147] W. A. Sauck. *A conceptual model for the geoelectrical response of LNAPL plumes in granular sediments*, in *Proceedings of the Symposium on the Application of Geophysics to Engineering and Environmental Problems: Chicago, Illinois, U.S.A.*, Environmental and Engineering Geophysical Society, 805-817, 1998.

[148] M. Simon. *Bayesian anomaly detection in heterogeneous media with applications to geophysical tomography*, Inverse Problems, **30** (2014), 114013.

[149] N. A. Simonov and M. Mascagni. *Random Walk Algorithms for Estimating Effective Properties of Digitized Porous Media*, Monte Carlo Meth. and Appl., **10** (2004), 599–608.

[150] E. Somersalo, M. Cheney and D. Isaacson. *Existence and uniqueness for electrode models for electric current computed tomography*, SIAM J. Appl. Math., **52** (1992), 1023–1040.

[151] D. W. Stroock. *Diffusion semigroups corresponding to uniformly elliptic divergence form operators*, in *Séminaire de Probabilités, XXII*, Springer, Berlin, 316–347, 1988.

[152] D. W. Stroock and S. R. S. Varadhan. *Diffusion processes with boundary conditions*, Comm. Pure Appl. Math., **24** (1971), 147–225.

[153] D. W. Stroock and S. R. S. Varadhan. *Multidimensional diffusion processes*, Springer-Verlag, Berlin, 2006.

[154] A. M. Stuart. *Inverse problems: a Bayesian perspective*, Acta Numer., **19** (2010), 451–559.

[155] J. Sylvester and G. Uhlmann. *A global uniqueness theorem for an inverse boundary value problem*, Ann. of Math., **125** (1987), 153–169.

[156] H. Tanaka. *Homogenization of diffusion processes with boundary conditions*, Stoc. Anal. Appl., Adv. Probab. Related Topics, **7** (1984), 411–437.

[157] S. Torquato. *Random heterogeneous media: microstructure and improved bounds on effective properties*, Appl. Mech. Rev., **44** (1991), 37–76.

[158] S. Torquato. *Random heterogeneous materials*, Springer-Verlag, New York, 2002.

[159] S. Torquato, I. C. Kim and D. Cule. *Effective conductivity, dielectric constant, and diffusion coefficient of digitized composite media via first-passage-time-equations*, J. Appl. Phys. **85** (1999), 1560–1571.

[160] G. M. Troianiello. *Elliptic differential equations and obstacle problems*, Plenum Press, New York, 1987.

[161] M. Vauhkonen, W. R. B. Lionheart, L. M. Heikkinen, P. J. Vauhkonen and J. P. Kaipio. *A MATLAB package for the EIDORS project to reconstruct two-dimensional EIT images* Physiol. Meas., **22** (2001), 107–111.

[162] K. Yosida. *Functional analysis*, Springer-Verlag, Berlin-New York, 1980.

[163] T. Zhang. *A probabilistic approach to Dirichlet problems of semilinear elliptic PDEs with singular coefficients*, Ann. Probab., **39** (2011), 1502–1527.

[164] V. V. Žikov, S. M. Kozlov, O. A. Oleĭnik and Ha T'en Ngoan. *Averaging and G-convergence of differential operators*, Uspekhi Mat. Nauk, **34** (1979), 65–133.

[165] V. V. Žikov, S. M. Kozlov, O. A. Oleĭnik. *Homogenization of differential operators and integral functionals*, Springer-Verlag, Berin, 1994.